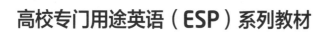

A PRACTICAL COURSE of
EST Translation

科技英语翻译实用教程

主　编　谢小苑

编　者　王珺琳　徐智鑫
　　　　王秀文　刘长江
　　　　何　烨　刘　莺
　　　　万　梅

清华大学出版社
北京

内容简介

本书系统介绍翻译的基础知识和科技英语翻译中词法、句法、章法和文体的特点及翻译技巧，通过各种科技文体的翻译实践，学生可掌握基本的翻译技能并达到一定的熟练程度。

本书共十二个单元，每个单元由三部分组成。第一部分"讲座"，主要讲解翻译基础知识，介绍科技英语翻译中词法、句法、章法和文体的主要特点与翻译技巧。第二部分"阅读"，文章内容突出科技特色，注释力求简明扼要，突出重点，讲清难点。第三部分"拓展阅读"，话题与第二部分相同，目的是给学生提供更多的翻译实践机会，同时，学生通过大量阅读增加词汇量，开阔视野。书后附有每单元的练习答案，便于学生核对和自主学习；所附课文译文，仅供对照参考。

本书适合作为高校非英语专业本科生的科技英语翻译教材，也可供广大科技人员和科技英语爱好者阅读参考。

版权所有，侵权必究。举报：010-62782989，beiqinquan@tup.tsinghua.edu.cn。

图书在版编目（CIP）数据

科技英语翻译实用教程/谢小苑主编. —北京：清华大学出版社，2020.4（2024.7 重印）
高校专门用途英语（ESP）系列教材
ISBN 978-7-302-55176-8

Ⅰ.①科… Ⅱ.①谢… Ⅲ.①科学技术—英语—翻译—高等学校—教材 Ⅳ.① N43

中国版本图书馆 CIP 数据核字（2020）第 049563 号

责任编辑：周 航 刘 艳
封面设计：子 一
责任校对：王凤芝
责任印制：宋 林

出版发行：清华大学出版社
    网 址：https://www.tup.com.cn，https://www.wqxuetang.com
    地 址：北京清华大学学研大厦 A 座 邮 编：100084
    社 总 机：010-83470000 邮 购：010-62786544
    投稿与读者服务：010-62776969，c-service@tup.tsinghua.edu.cn
    质量反馈：010-62772015，zhiliang@tup.tsinghua.edu.cn
印 装 者：小森印刷霸州有限公司
经 销：全国新华书店
开 本：170mm×230mm 印 张：22.5 字 数：366 千字
版 次：2020 年 6 月第 1 版 印 次：2024 年 7 月第 7 次印刷
定 价：75.00 元

产品编号：086749-02

前言

随着世界经济全球化、科技现代化、文化多元化进程的加速,国家和社会对非英语专业学生的英语应用能力提出了新的要求,对大学英语教学提出了新的挑战。新出台的《大学英语教学指南》明确要求大学英语课程教学内容从单一的通用英语向"通用英语+"型,即通用英语+专门用途英语(ESP)和跨文化交际的一体两翼转变(王守仁,2016)。为适应新形势,满足新时期国家和社会对人才培养的需求,南京航空航天大学加强专门用途英语类和跨文化交际类课程建设,突出知识构建、能力培养和文化素质的提高,使学生的英语知识、能力、素质得到协调发展,培养具有责任意识、创新精神、国际视野和人文情怀的社会栋梁和工程英才,实现新形势下大学英语教学的可持续发展,促进大学英语课程的长远发展。

《科技英语翻译实用教程》和《科技英语阅读实用教程》是南京航空航天大学大学英语教学改革探索和专门用途英语教学团队教学实践积累的成果,旨在帮助学生顺利阅读和翻译所学专业的英语文献和资料。我们期待该系列教材能为学生知识、能力、素质的协调发展和英语应用能力的提升提供一个良好的契机和新的生长点。

《科技英语翻译实用教程》具有以下特色:

1. 技巧与实践相结合

本书在编排上力求有所创新,突显技巧与实践的紧密结合。本书以"讲座"的形式讲解科技英语翻译中词法、句法、章法和文体的翻译技巧,以各类科技文体的翻译作为实践内容。翻译技巧"讲座"既体现科技英语翻译的规律和方法,又提供大量的例句和相应的翻译实践,做到讲练结合;"课文"及"拓展阅读"的翻译实践以翻译理论与技巧为指导,做到学用结合。

2. 知识性与实用性相结合

本书的编写注重实用,内容力求做到深入浅出、通俗易懂,目的是帮助学生掌握科技英语翻译的基础知识和技能。本书所选课文、译例和练习内容涉及科学与技术的许多学科领域，如物理科学、地球科学、生命科学、航空、航天、民航、计算机、机械工程等，学生在学习翻译技巧与进行翻译实践的同时，也可以了解许多相关的学科知识，扩大科技英语的词汇量，开阔视野，提高科学文化素养。

3. 系统性与针对性相结合

本书编者认为，一本较好的科技英语翻译教材应该研究学习者本身的特点和与之相关联的工作、社会、未来等的需求，应该结合学习者所学专业及相关学科。因此，本书针对高校大学生（尤其是理工科学生）的特点，按照用人单位的需求，系统介绍翻译的基础知识、科技英语翻译各层次的特点及翻译技巧，重点讲解学生在翻译中经常碰到的各种问题及解决方法，选择学生未来工作中会经常接触到的文体形式及来源于实际运用中的语言材料，帮助学生尽快了解自己感兴趣的领域，以适应实际工作和社会的需要。

本书共十二个单元，每个单元由三部分组成。第一部分"讲座"，主要讲解翻译基础知识，介绍科技英语翻译中词法、句法、章法和文体的主要特点与翻译技巧；每一讲之后的"即学即测"，精选了内容丰富、形式多样的练习，巩固学生已学的知识。第二部分"阅读"，文章内容突出科技特色，注释力求简明扼要，突出重点，讲清难点。第三部分"拓展阅读"，话题与第二部分相同，目的是给学生提供更多的翻译实践机会，同时，学生通过大量阅读增加词汇量，开阔视野。书后附有每单元的练习答案，便于学生核对和自主学习；所附课文译文，仅供对照参考。

本书得以付梓，离不开方方面面的支持。首先，本书为南京航空航天大学2019年校级教育教学改革项目（精品教材建设专项）的研究成果，我们对学校项目的资助表示感谢。其次，感谢南京航空航天大学的各级领导，尤其是教务处及教材科、外国语学院领导的支持，他们在经费和政策上的大力支持为本书的顺利完成提供了有力保障。再次，感谢参与教材编写与修订的专门用途英语教学团队成员，他们基于《科技英语翻译技巧与实践》（2008）和《科技英语翻译》（2015），对本书内容进行补充和更新，使其更具南京航空航天大学的航空、航天、民航特色。本书在编写过程中，参考了国内外出版的

前　言

相关书刊并引用了部分资料，在此向有关作者和单位表示诚挚的感谢。

由于编者水平和经验有限，书中欠妥与谬误之处在所难免，祈请同行专家和广大读者，不吝批评，多多斧正，以便今后修订完善。

编者

2019 年 10 月于南京航空航天大学

Contents
目 录

Unit One
Science and Technology 1
- **Part A**   **Lecture**   科技英语的特点 2
- **Part B**   **Reading**   Pure and Applied Science 10
- **Part C**   **Extended Reading**   Branches of Science 14

Unit Two
Science and Technology in the Past 17
- **Part A**   **Lecture**   翻译概论 18
- **Part B**   **Reading**   Science and Technology in Traditional China 35
- **Part C**   **Extended Reading**   List of Chinese Inventions 38

Unit Three
Language of Science 43
- **Part A**   **Lecture**   科技术语的翻译 44
- **Part B**   **Reading**   Numbers and Mathematics 52
- **Part C**   **Extended Reading**   Mathematical Methods in Physics and Engineering 55

Unit Four
Physical Science 59
- **Part A**   **Lecture**   词类转换法 60

Part B   **Reading**   Physics 75
Part C   **Extended Reading**   Relationship of Chemistry to Other
                                       Sciences and Industry 80

Unit Five
Earth Science 83

Part A   **Lecture**   增词法与减词法 84
Part B   **Reading**   The Scope of Geology 102
Part C   **Extended Reading**   Paleontology 106

Unit Six
Life Science (I) 109

Part A   **Lecture**   被动语态的翻译 110
Part B   **Reading**   History of Biology 124
Part C   **Extended Reading**   Branches of Biology 129

Unit Seven
Life Science (II) 135

Part A   **Lecture**   定语从句的翻译 136
Part B   **Reading**   It's Not "All in the Genes" 156
Part C   **Extended Reading**   Genetically Modified Foods 161

Unit Eight
Aeronautics 165

Part A   **Lecture**   长句的翻译 166
Part B   **Reading**   The Plane Makers 176
Part C   **Extended Reading**   How Aircraft Are Built 180

目 录

Unit Nine
Astronautics **185**

    Part A   Lecture 段落的翻译 186
    Part B   Reading The Scientific Exploration of Space 199
    Part C   Extended Reading Introduction to Space Exploration 203

Unit Ten
Civil Aviation **207**

    Part A   Lecture 科技文章的翻译 208
    Part B   Reading Civil Aviation Faces Green Challenge 227
    Part C   Extended Reading A Brief Introduction to the International Public Air Law 233

Unit Eleven
Computer Science **237**

    Part A   Lecture 产品说明书的翻译 238
    Part B   Reading Artificial Intelligence 251
    Part C   Extended Reading Big Data 255

Unit Twelve
Mechanical Engineering **261**

    Part A   Lecture 科技文章检索 262
    Part B   Reading The Engineering Profession 274
    Part C   Extended Reading Functions of Mechanical Engineering 279

References **283**
Key and Reference for Translation **287**

Unit One
Science and Technology

Part A Lecture

科技英语的特点

　　随着世界科技发展的日新月异和我国对外科技交流的日益频繁，科技英语翻译的重要性越来越清楚地显现出来。为了更好地促进国民经济发展、加强科技交流，同时也为了提高自身的专业素质，大学生需要学习并掌握科技英语翻译方面的知识。本书主要讨论科技英语翻译技巧，但在谈技巧之前先介绍科技英语及其特点。

1 科技英语概述

　　科技英语是一种重要的英语语体，也称科技文体。它是随着科学技术的发展而形成的一种独立的文体形式。科技英语既涵盖自然科学领域的各种知识和技术，也包括社会科学的各个领域，如用英语撰写的有关自然科学和社会科学的学术著作、论文、实验报告、专利、产品说明书等。

　　科技英语有别于通用英语。自 20 世纪 70 年代以来，科技英语就引发了国际上的广泛关注，并在教育领域成为一门专业，科技英语的重要性日益突显。有人认为，只要懂英语语法和一些科技词汇，就能理解科技英语，即科技英语 = 英语语法 + 科技词汇。其实，科技英语并不像他们想象的那样简单。科技英语在词汇、语法和文体上都具有自己的特点。

2 科技英语的特点

　　为了能准确、简洁明了地叙述自然现象、事实及其发展过程、性质和特征，科技人员喜欢在文章中使用一些典型的句型和大量的专业术语，因而形成科技英语自身的特色。本节主要通过列举实例的方法阐明科技英语在词汇、语法和文体上的特点。

Unit One Science and Technology

2.1 词汇特点

大量使用科技术语是科技英语的基本特点，这是因为科技方面的专业术语是构成科技理论的语言基础，其语义单一且严谨。为了概括社会科学和自然科学等方面的现象，揭示客观事物的发展规律，科技英语必须使用表意明确的专业术语。

Some of the most common methods of inputting information are to use *magnetic tape, disks,* and *terminals.* The computer's *input device* (which might be a *key-board,* a *tape drive* or *disk drive,* depending on the *medium* used in inputting information) reads the information into the computer. For outputting information, two common devices used are a *printer* which prints the new information on paper, and a *CRT display screen* which shows the results on a TV-like screen.

输入信息的一些最普通的方法是使用**磁带**、**磁盘**和**终端**。计算机的**输入装置**（依据输入信息时使用的**媒体**，输入装置可能是**键盘**、**磁带机**或**磁盘驱动器**）把信息读入计算机内。对于输出信息，有两种常用的装置：把新信息打印在纸上的**打印机**，以及在类似电视的荧屏上显示结果的**阴极射线管显示屏**。

2.2 语法特点

科技英语在词法和句法的运用上和通用英语不同，词法上主要表现在大量使用名词化结构，句法上主要表现在时态的不同用法、广泛使用被动语态、大量使用非限定动词和大量使用长复句。

▶ 2.2.1 大量使用名词化结构

科技英语在词法上的显著特点是大量使用名词化结构。大量使用名词化结构主要指广泛使用能表示动作或状态的抽象名词或起名词作用的非限定动词。科技文章的任务是叙述事实和论证推断，因而科技文体要求行文简洁、表达客观、内容确切、信息量大，大量使用名词化结构正好符合科技文体的要求。

例2

Archimedes first discovered the principle *of displacement of water by solid bodies*.

(=Archimedes first discovered the principle that water is displaced by solid bodies.)

阿基米德最先发现固体排水的原理。

2.2.2 时态的不同用法

尽管科技英语中常用的时态有一般现在时、一般过去时和现在完成时，但一般现在时是最常用的时态。在大多数的科技文章中，科技人员会使用"无时间性"的一般现在时。这是因为科技书籍包含关于科学知识的现状、关于科学知识的各种实验以及如何利用这些知识的信息。

在科技英语中，一般过去时和现在完成时这两种过去时态常用在科技发展史、科技报告和科技报纸杂志（即有关科学和科学家的新闻报道）中。

An experiment to measure atmospheric pressure (after Torricelli)

First, a long glass tube *is* taken. The tube *is* closed at the top and *is* then completely filled with water. Next it *is* placed vertically in a large barrel half-full of water. When the bottom of the tube *is* opened, the water level in the tube only *falls* to a height of approximately 10 meters above the water level in the barrel. As a result, a vacuum *is* left in the upper part of the tube. The water in the tube *is* supported by the atmospheric pressure. The height of the column of water *can* therefore be used to measure atmospheric pressure.

测量大气压力的实验（仿照托里拆利）

首先取一根长玻璃管，将顶端封闭并盛满水，然后竖直地放在一只水半满的大桶中。当管的底部打开时，管中的水面只下降到大桶水面之上大约10米高度处。结果，在管的上部就会留下真空。管内的水为大气压力所支撑，因此水柱的高度可用来测量大气压力。

2.2.3 广泛使用被动语态

根据英国利兹大学 John Swales 的统计，科技英语中大概有三分之一的动词是被动语态。这是因为科技英语叙述的往往是客体，即客观的事物、现象或过程，而主体往往是从事实验、研究和分析的人或装置。使用被动语态比较客观，还能使读者的注意力集中在客体上。

For this reason, computers ***can be defined*** as very-high-speed electronic devices which accept information in the form of instructions called a program and characters called data, perform mathematical and/or logical operations on the information, and then supply results of these operations.

因此，可以把计算机**定义为**一种高速运作的电子设备。它以指令（称为程序）和字符（称为数据）的形式接收信息，并对这些信息执行数学的和（或）逻辑的操作，然后提供这些操作的结果。

Computers ***are thought to*** have many remarkable powers.

人们认为计算机有许多神奇的功能。

2.2.4 大量使用非限定动词

科技文章要求语言简练，结构紧凑。因此，科技英语中大量使用非限定动词，即分词、不定式和动名词，特别是分词。

The computer's input device (which might be a keyboard, a tape drive or disk drive, ***depending on*** the medium ***used*** in inputting information) reads the information into the computer.

计算机的输入装置（**依据**输入信息时**使用的**媒体，可能是键盘、磁带机或磁盘驱动器）把信息读入计算机内。

▶ 2.2.5 大量使用长复句

科技文章逻辑严密，结构紧凑，因此，科技英语中往往出现许多长句。长句一般有两种：一种是带有较多定语和状语的简单句，一种是包含多个从句（如定语从句、状语从句）或分句的复合句与并列复合句。

A computer cannot do anything unless a person tells it what to do and gives it the appropriate information; *but* because electric pulses can move at the speed of light, a computer can carry out vast numbers of arithmetic-logical operations almost instantaneously.

计算机不能做任何事情，除非人们告诉它做什么并且给它一些恰当的信息；**但是**因为电子脉冲能够以光速运动，所以计算机能够瞬间执行大量算术—逻辑运算。

科技英语在语法上的特点还表现在文章中常出现表示逻辑关系的连接词，这是因为科技英语重视叙事的逻辑性、层次感和转折、对比以及推出前提、列出条件、导出结论等论证手段。

Computers are thought to have many remarkable powers. *However*, most computers, whether large or small, have three basic capabilities. *First*, computers have circuits for performing arithmetic operations, such as: addition, subtraction, multiplication, division and exponentiation. *Second*, computers have a means of communicating with the user. After all, if we couldn't feed information in and get results back, these machines wouldn't be of much use. *However*, certain computers (commonly minicomputers and microcomputers) are used to directly control things such as robots, aircraft navigation systems, medical instruments, etc. *Third*, computers have circuits which can make decisions. The kinds of decisions which computer circuits can make are not of the type: "Who would win a war between two countries?" or "Who is the richest person in the world?" Unfortunately, the computer can only decide three things, namely: Is one number less than another? Are two numbers equal? And, is one number greater than another?

Unit One   Science and Technology

　　人们认为计算机有许多神奇的功能。**然而**，大多数计算机，无论大小，都具有三种基本功能。**第一**，计算机有执行算术运算的电路系统，如加、减、乘、除和取幂。**第二**，计算机有与用户交流的方法。毕竟，如果我们不能输入信息并取回结果，这些机器也就不会有太大用处。**然而**，某些计算机（通常是小型计算机和微型计算机）被用来直接控制物体，如机器人、飞机导航系统、医疗设备等。**第三**，计算机有能够做出判断的电路系统。遗憾的是，计算机系统做不出"两个国家谁将赢得这场战争？"或"谁是世界上最富有的人？"这样的判定。计算机只能判断三件事：一个数是否小于另一个数，两个数是否相等，以及一个数是否大于另一个数。

2·3 文体特点

　　科技英语非常注重逻辑上的连贯，思维上的准确和严密，表达上的清晰与精练，以客观的风格陈述事实和揭示真理。因此，科技英语避免表露个人感情，力求少用或不用充满感情色彩的词，尽量避免使用旨在加强语言感染力的各种修辞格。

　　The general layout of the illumination system and lenses of the electron microscope corresponds to the layout of the light microscope. The electron "gun" which produces the electrons is equivalent to the light source of the optical microscope. The electrons are accelerated by a high-voltage potential (usually 40,000 to 100,000 volts), and pass through a condenser lens system usually composed of two magnetic lenses. The system concentrates the beam onto the specimen, and the objective lens provides the primary magnification. The final images in the electron microscope must be projected onto a phosphor-coated screen so that it can be seen. For this reason, the lenses that are equivalent of the eyepiece in an optical microscope are called "projector" lenses.

　　电子显微镜的聚光系统和透镜的总体设计与光学显微镜的设计是一致的。产生电子束的电子"枪"相当于光学显微镜的光源。电子被高压（通常为40,000伏—100,000伏）的电位差加速，穿过聚光镜系统。聚光镜通常由两组磁透镜组成；聚光镜系统可将电子束聚集在样品上，并由物镜提供主要的放大倍率。电子显微镜的最终成像被投射到磷光屏上，以便进行观察。正是由于这个原因，这些相当于光学显微镜目镜的透镜被称为"投影镜"。

通过以上讲解，我们对科技英语在词汇、语法和文体上的特点有了一个比较全面的认识，但要真正翻译好科技文章，还需进一步了解科技翻译的特点以及各种翻译技巧，还需进行大量的翻译实践。

Quiz

1. Translate the following sentences into Chinese. While translating, pay more attention to the features of EST.

    (1) The substitution of some rolling friction for sliding friction results in a very considerable reduction in friction.

    (2) Nature rubber is obtained from rubber trees as a white, milky liquid known as latex. This is treated with acid and dried, before being dispatched to countries all over the world.

    (3) Today the electronic computer is widely used in solving mathematical problems having to do with weather forecasting and putting satellite into orbit.

    (4) When steam is condensed again to water, the same amount of heat is given out as it was taken in when the steam was formed.

    (5) In radiation, thermal energy is transformed into radiant energy, similar in nature to light.

    (6) This position was completely reversed by Haber's development of the utilization of nitrogen from the air.

    (7) Two-eyed, present-day man has no need of such microscopic delicacy in his vision.

    (8) This is an electrical method, which is most promising when the water is brackish.

Unit One　Science and Technology

(9) It was understood that atoms were the smallest elements. It is known now that atoms are further divided into nuclei and electrons, neutrons and protons, etc.

(10) If there had not been any air in the cooling system, the effect of cooling would not have been affected and the temperature could not have been kept so low.

(11) Experiments show that there is a definite relationship among the electrical pressure that makes a current flow, the rate at which the electricity flows and the resistance of the object or objects through which the current passes.

(12) The efforts that have been made to explain optical phenomena by means of the hypothesis of a medium having the same physical character as an elastic solid body led, in the first instance, to the understanding of a concrete example of a medium which can transmit transverse vibration but later to the definite conclusion that there is no luminiferous (发光的) medium having the physical character assumed in the hypothesis.

Part B Reading

Pure and Applied Science

As students of science[1], you are probably sometimes puzzled by the terms "pure" and "applied" science. Are these two totally different activities, having little or no interconnection, as is often implied? Let us begin by examining what is done by each.[2]

Pure science **is** primarily **concerned with** the development of theories (or, as they are frequently called, models) establishing relationships between the phenomena of the universe.[3] When they are sufficiently **validated**, these theories (hypotheses, models) become the working laws or principles of science. In carrying out this work, the pure scientist usually **disregards** its application to practical affairs, **confining** his attention to explanations of how and why events occur. Hence, in physics, the equations describing the behavior of fundamental particles, or in biology, the establishment of the life cycle of a particular species of insect living in a Polar environment[4], are said to be examples of pure science (basic research), having no **apparent** connection (for the moment) with technology, i.e., applied science.

Applied science, on the other hand, is directly concerned with the application of the working laws of pure science to the practical affairs of life, and to increasing one's control over his environment, thus leading to the development of new techniques, processes and machines.[5] Such activities as investigating the strength and uses of materials, extending the findings of pure mathematics to improve the sampling procedures used in agriculture or the social sciences, and developing the **potentialities** of atomic energy, are all examples of the work of the applied scientist or technologist.

It is evident that many branches of applied science are practical extensions of purely theoretical or experimental work. Thus the study of radioactivity began as a piece of pure research, but its results are now applied in a great number of different ways—in the cancer treatment in medicine, the development of **fertilizers** in agriculture, the study of metal-fatigue in engineering, the methods of estimating the ages of objects in anthropology and geology, etc. **Conversely**, work in applied science and technology frequently acts as a direct **stimulus** to the development of pure science. Such an interaction occurs, for example, when the technologist, in applying a particular concept of pure science to a practical problem, reveals a gap or limitation in the theoretical model, thus pointing the way for further basic research. Often a further interaction occurs, since the pure scientist is unable to undertake this further research until another technologist provides him with more highly-developed instruments.

It seems, then[6], that these two branches of science are **mutually** dependent and interacting, and that the so-called **division**[7] between the pure scientist and the applied scientist is more apparent than real.

New Words and Expressions

be concerned with 涉及	fertilizer 肥料
validate 有效	conversely 相反地，另一方面
disregard 不顾，不理	stimulus 刺激（物），促进因素
confine 限制	mutually 相互地
apparent 明显的	division 区分
potentiality 潜力	

Notes

1. students of science 指理科学生，本文指学习自然科学的学生，也可译为理工科学生。
2. 句中 examine 意为检查、审查、考查。全句直译为：让我们从考查各自所做的开始吧。

3. 句中 establishing 引起的分词短语，是修饰 theories 的。全句直译为：理论科学涉及的是确立把宇宙间的种种现象联系起来的理论（或者人们通常所称的模型）。

4. Polar 指的是南极和北极，Polar environment 说的是南北两极极其寒冷的环境。

5. 句中 the development of new techniques, processes and machines，直译是"发展新技术、新工艺、新机器"，但翻译成汉语时 development 采用了三种不同的搭配——新技术的发展、新工艺的制作和新机器的研制。

6. then 在这里有承接前文、总结全文的意义。

7. so-called division 指形式上的区分，real 是 real division，指实质上的区分。

Exercises

1. Find out the English equivalents of the following Chinese terms from the passage.

   （1）理论科学　　　　　　（2）应用科学
   （3）模型　　　　　　　　（4）假设
   （5）现行的科学定律　　　（6）生命周期
   （7）方程式　　　　　　　（8）放射性
   （9）金属疲劳　　　　　　（10）人类学

2. Translate the following sentences into Chinese.

   (1) Pure science is primarily concerned with the development of theories (or, as they are frequently called, models) establishing relationships between the phenomena of the universe.

   (2) In carrying out this work, the pure scientist usually disregards its application to practical affairs, confining his attention to

explanations of how and why events occur.

(3) Such activities as investigating the strength and uses of materials, extending the findings of pure mathematics to improve the sampling procedures used in agriculture or the social sciences, and developing the potentialities of atomic energy, are all examples of the work of the applied scientist or technologist.

(4) It is evident that many branches of applied science are practical extensions of purely theoretical or experimental work.

(5) It seems, then, that these two branches of science are mutually dependent and interacting, and that the so-called division between the pure scientist and the applied scientist is more apparent than real.

Part C Extended Reading

Branches of Science

Science may be roughly divided into the physical sciences, the earth sciences, and the life sciences. Mathematics, while not a science, **is** closely **allied to** the sciences because of their extensive use of it. Indeed, it **is** frequently **referred to as** the language of science, the most important and objective means for communicating the results of science. The physical sciences include physics, chemistry and astronomy; the earth sciences (sometimes considered a part of the physical sciences) include geology, paleontology, oceanography, and meteorology; and the life sciences include all the branches of biology such as botany, zoology, genetics, and medicine. Each of these subjects is itself divided into different branches, e.g., mathematics into **arithmetic, algebra, geometry,** and **analysis**; physics into mechanics, thermodynamics, **optics, acoustics,** electricity and magnetism, and atomic and nuclear physics. In addition to these separate branches, there are numerous fields that **draw on** more than one branch of science, e.g., astrophysics, biophysics, biochemistry, geochemistry, and geophysics.

All of these areas of study might be called pure science, **in contrast to** the applied or engineering, science, i.e., technology, which is concerned with the practical application of the results of scientific activity. Such fields include mechanical, civil, **aeronautical**, electrical, architectural, chemical, and other kinds of engineering; **agronomy**, horticulture, and animal husbandry; and many aspects of medicine. Finally, there are distinct disciplines for the study of the history and philosophy of science.

Unit One Science and Technology

New Words and Expressions

be allied to 与……有关联
be referred to as（某人/某物）被称为
arithmetic 算术
algebra 代数学
geometry 几何学
analysis 解析学，分析学
optics 光学
acoustics 声学
draw on 利用，凭，靠
in contrast to 与……形成对照
aeronautical 航空的
agronomy 农学，农艺学，作物学

Exercises

1. **Find out the English equivalents of the following Chinese terms from the passage.**

   （1）物理科学 （2）地球科学
   （3）生命科学 （4）天文学
   （5）古生物学 （6）海洋学
   （7）气象学   （8）植物学
   （9）动物学   （10）遗传学
   （11）力学/机械学 （12）热动力学
   （13）天文物理学 （14）园艺（学）
   （15）畜牧业

2. **Translate the following sentences into Chinese.**

   (1) Indeed, it is frequently referred to as the language of science, the most important and objective means for communicating the results of science.

   (2) In addition to these separate branches, there are numerous fields that draw on more than one branch of science, e.g., astrophysics, biophysics, biochemistry, geochemistry, and geophysics.

   (3) All of these areas of study might be called pure science, in contrast to the applied or engineering, science, i.e., technology, which is concerned with the practical application of the results of scientific activity.

Unit Two

Science and Technology in the Past

Part A Lecture

翻译概论

翻译的渊源,可以上溯到几千年前的古代社会。当使用不同语言的人们由于政治、经济、文化、宗教等原因开始相互交往时,翻译也就应运而生。这些翻译活动使得操持不同语言的民族之间得以交流思想和相互往来,促进了人类文明的发展和进步。随着经济全球化和教育国际化进程的加速,新的形势迫切要求大学生加强翻译理论与技巧的研究与学习,提高翻译质量。下面将对翻译的一些基础知识分别予以介绍。

1 翻译的目的

语言是人类区别于其他动物的最主要特征,但是不同国家、不同民族由于发展历史不同和生活区域不同,在表达思想感情时所使用的语言各不相同,如汉语、日语、韩语、英语、法语、德语、西班牙语等。使用不同语言的人们要进行思想情感的交流,就会遇到语言方面的障碍。例如,一个没有学过外语的中国人,当他遇见日本人、英国人、法国人或者其他外国人时,依靠语言是不可能得以交流沟通的。一个工厂花大笔资金购买国外先进设备,若没有对这些设备的口头或书面翻译,再先进的设备,也只是一堆破铜烂铁,不能发挥应有的作用。当然,这种语言方面的障碍可以通过一个懂外语的人的口头或书面翻译来克服,因为一种语言所表达的思维内容完全可以用另一种语言重新表达出来(**参见本章 2.1**)。改变其声,以传达其意,是为口译;改变其形,以传达其意,是为笔译。随着科学技术的迅猛发展,语言方面的障碍甚至可以借助机器来实现。

翻译是沟通各族人民的思想和促进政治、经济、文化、科学、技术交流的重要手段。当使用不同语言的民族之间进行交往时,必然要以翻译作为桥梁。没有翻译,人们彼此说的话互不理解,思想就难以沟通,交往就无法进行。在经济全球化的今天,随着我国逐渐融入世贸大家庭以及改革开放程度的不断深入,我们与世界各国的交往日益频繁,翻译工作的重要性也与日俱增。

Unit Two　Science and Technology in the Past

　　翻译也是学习外语的重要手段之一。一篇课文学过之后，并不一定能够彻底了解。一般来说，阅读课文只限于理解句、段、篇章的大意，而翻译则要透彻了解其中每个单词、词组的确切含义，因此，翻译能够帮助我们加深对原文的理解，提高阅读能力。翻译也是语言学习所要掌握的听、说、读、写、译五大技能之一。学习翻译还能帮助我们练习中文写作，学会推理思考的方法，养成细心、耐心、认真负责的工作态度。翻译还是探讨两种语言对应关系的一门学科。通过大量的翻译实践，我们能更加准确地了解和掌握两种语言之间的异同，通过比较和对比，使我们更加快捷地学会地道的外语。

　　总之，翻译本身不仅是一种技能，它还是一门工具、一个手段，能够帮助我们发展其他方面的能力。"翻译"一词实际上有两重含义，既指翻译的过程，又指这一过程所产生的成品，即译文。翻译的过程同时就是理解、学习和交流的过程，翻译的成品也是为了帮助我们克服语言的障碍去学习了解外国的风俗文化等。因此，"过程"和"成品"对我们都是有重要意义的，都是我们翻译的目的。我们既要重视翻译的过程，同时也要重视翻译活动的结果——译文。翻译就是为了交流、学习和提高。

　　科技英语翻译就是为了让使用不同语言的人摆脱英语方面的障碍，准确地了解、学习和交流彼此在科学技术领域所取得的既有成就和最新成果。随着信息时代的到来，对外交流的日益频繁，科学技术突飞猛进的发展，科技翻译正扮演着一个越来越重要的角色，对社会发展的影响也日益深刻。任何个人或企业要想走上国际舞台，寻求更广阔的发展空间，常常得借助于翻译工作，尤其是科技翻译。翻译是架设于不同国家、不同民族的人们之间的一座金桥。

2 翻译的性质

　　为了说明翻译的性质，本节试图从语言的可译性与不可译性、翻译的定义、翻译到底是一门艺术还是一门科学等三个方面来分别予以阐述，以便大家能对翻译的性质有一个较全面的了解。

2.1 可译性与不可译性

可译性与不可译性的争论由来已久。语言理论界对语言是否可译，翻译是否可能，特别是对语际翻译是否可能的问题，存在着针锋相对的见解。一种观点认为，语言的深层结构是普遍存在的，而且是共同的，不管是什么语言，它们都有共同的深层结构，差异主要在于表层，因此翻译是可能的，因为所有的思想观念都可以用人类现有的任何一种语言来表达。进行翻译就是要透过两种语言的表层，使处于深层的共同的东西充分发挥作用。对立的观点则认为，所谓普遍存在的深层结构属于逻辑和心理的因素，显得过于抽象笼统而无法把握，我们所面对的不同语言在词法、句法、文化底蕴等方面都各不相同，因而在不同的语言之间真正的翻译是不可能的，所谓翻译只不过是粗糙的仿制品或近似物。

现代语言学理论的发展在这一问题上给我们带来了一些深刻的启示。当代语言学认为，虽然使用不同语言的人们进行交流时会遇到困难和障碍，但由于宇宙共相、生态共相、生理共相和心理共相的存在，必然导致语言共相或语言之间的相似性，这就为不同语言间的交流和翻译提供了较为充足的基础。语言是思维的外壳，人类的思维是由存在决定的，而客观存在是可以认识的，因此一种语言所表达的思维内容，用另一种语言重新表达出来是完全可能的。

当然，我们也要意识到，这种语际的转换是有一定限度的，因为翻译时有时会出现一些语言不可译或文化不可译现象。如汉语歇后语"孔夫子搬家——尽是书（输）"中，利用"书""输"之间的谐音进行转义，若翻译成英语，就很难做到十全十美。一些双关语、文化负载词语等的翻译也是如此，比如像"三顾茅庐""桃园结义"这种蕴涵文化典故的成语的翻译就很难用简洁紧凑的语言传输原文所蕴涵的内容。将英语里的一些成语和谚语汉译时，也常常是形式和内容难以兼顾，很难做到尽善尽美。

但不管怎样，人类的共性、交流的必要性以及语言本身的共性，使得翻译是可取的，是可行的，也是必须的。虽然翻译总有一定的局限性，难以做到完美，但我们总是要追求尽善尽美，追求最大程度的可译性。当形式和内容难以同时兼顾时，我们常常利用各种翻译技巧，采用意译的方式来实现语际信息传输。

2·2 翻译的定义

西方许多翻译理论家在阐明自己的观点之前都对翻译下了精确的定义，如英国著名的语言学家和翻译理论家卡特·福德将翻译定义为"用译语中对等的语言材料来替换原语中的语言材料"；苏联语言学派的代表人物之一巴尔胡达罗夫指出，翻译是把一种语言的话语转换为另一种语言的话语的动态过程；而在著名的美国语言学家尤金·奈达看来，"翻译是指接受语复制原语的信息的最近似的自然等值，首先在意义方面，其次在文体方面"，等等。

这些翻译理论家的表述各不相同，侧重点也有所差异，但都指出了翻译是把两种语言进行转换的过程。在我们看来，通俗地说，翻译是把一种语言所表达的思维内容、感情、风格等用另一种语言表现出来的语言活动。这种语言活动是一种创造性的活动，因为一篇好的译文不仅需要如实地表达原文的思想，忠实于原文，而且还应将原文的文字技巧、写作风格在译文中再现出来，达到既能表意又能传神的境界。要做到这些，就需要译者对原文的内容、风格等进行全面正确的理解，然后用最贴切最恰当的对应语将其表述出来。

科技英语翻译，顾名思义，主要是指将那些用英语写成的科技文章或材料转换成用规范汉语表述的文章或材料的语言活动。科技英语翻译是翻译的一种，理应遵循翻译的一般规律和技巧，但由于科技文体有其自身的规范和特点，科技英语翻译也有一些自身的特点和技巧。对此，后文将详细表述（**参见本章第 4 节**）。

2·3 科学还是艺术

翻译是一门科学还是一门艺术，人们对于这个问题也长期争论不休。

艺术表现个性，折射着创作者的独特思维与视角，所以艺术创作千差万别，各具特色，即使同一题材、同一内容，每个人的表现方法也各不相同。同一作品对欣赏者所产生的效果不尽相同。与艺术不同，科学则表现共性，不容变动。科学研究一般都按一定的规律，遵循一定的方法，得出相同的结果。

对于翻译来说，同样的原文，虽然每个译者都竭力保持原作的风格，不把自己的风格强加于原作者，力求让译文最大限度地忠实于原文，但每篇译文仍不可避免各显特色，所以有人坚持认为翻译是一门艺术而不是科学，它不能像科学一样制定出明确的规则和公式，而是依靠创造性的技巧运用。从这个意义上说，翻译在很大程度上是一种再创造。

相反，以德国应用语言学家威尔斯为代表的翻译科学论，基本上建立在一种不能由经验加以证实的有关语言本质的主张之上。就方法论而言，翻译科学派的理论家大都强调语言的普遍化与一般化，甚至到了将语言表达中那些独特的、不同的或新的东西统统剔除的程度，力图找到一些一般规律和科学方法来进行语言翻译研究和教学。所以，在他们看来，翻译是一门科学。

法国新时期著名的语言学家和翻译理论家乔治·穆南认为，翻译应该是一门艺术，是一门建立在科学基础上的艺术。我们认为，这个观点是对翻译本质的比较全面的认识。翻译是从一个符号系统转换成另一个符号系统的过程，在这个过程中，虽然译者总想尽量保持原作的风貌，忠实于原文，但译者的译文最终取决于他对原作的理解程度和他的目标语水平。虽然翻译活动中有很多科学规律和技巧，但不可否认，翻译的过程在某种程度上是一个再创造的过程。所以我们认为，翻译是一门建立在科学基础上的艺术。

对于科技翻译来说，由于科技文章以准确传输信息为主，其规律性较之文学翻译更强一些，可供艺术创作的空间相对要小一些，但是译文仍然会因译者对原文的理解程度不同和他的译语水平不同而有所差异。从这个意义上说，科技翻译仍然是一门建立在科学基础上的艺术。

3 翻译的标准

翻译的标准是指导翻译实践的准则，是衡量译文优劣的尺度。我国古代翻译界的巨星玄奘提出的翻译标准"既须求真，又须喻俗"，意即"忠实、通顺"，直到今天仍然具有指导意义。一百多年前，我国清末新兴资产阶级的启蒙思想家严复，参照古代佛经翻译的经验，根据自己翻译的实践，提出了著名的"信、达、雅"翻译标准。"信"是指"意义不倍（背）本文"，"达"是指不拘泥于原文形式，尽译文语言的能事以求原意明显，"雅"是指追求译文的古雅（这个标准受到很多人的批评，现在，这个标准常常被赋予新的含义，如"美"等）。鲁迅先生曾提出"信和顺"的翻译标准。他说："凡是翻译，必须兼顾着两面，一当然力求其易解，二则保存着原作的风姿。"关于翻译的标准，还有很多不同的主张和见解，如"信、达、切""达意、传神、文采""忠实、通顺、美""准确、流畅""神似""化境""明确、通顺、简练"等。

在实际工作中，我们认为，翻译一般要做到忠实和通顺，就是译文要力求确切地表达原文的内容和风格，同时在形式上又符合汉语的规范，做到通顺、流畅、易懂。忠实是通顺的基础，译文如果不忠实，再通顺也失去了意义。反过来，译文如果不通顺又必然影响到译文的忠实。所以一篇好的译文一定要做到忠实与通顺相结合，即用规范的译文语言形式，准确完整地表达原文的意思，这是我们对译文质量的基本要求。要达到这个基本要求就必须抓住原文的真正含义，但不要过多地受原文结构形式的束缚，而是按照汉语习惯用法来安排译文。当然，在不影响忠实和通顺的前提下，若能兼顾原文和译文形式上的统一，则更为理想。

在说到"通顺"这个标准时，应该补充一下的是，如果原文中出现故意的不通顺（多出现于文学作品中，如描写一个结巴的话语或一个没受过教育而词不达意的人的话语等），译文也应本着"忠实"的原则，故意译成不通顺，此所谓等效原则。等效原则就是指"从语义到文体在译语中用最近似的自然对等值再现原语的信息"，力争既传达信息，又传达原作的精神和风格，语言顺畅自然，使读者阅读译文和原文时反应类似；换句话说，就是翻译者应力争使译文对读者产生与原作一样的效果。

在科技文章中，上述故意的不通顺是不会出现的。对于科技翻译来说，很多人倾向于采用鲁迅先生的"信和顺"的翻译标准。根据科技翻译的性质和目的，我们认为，其质量标准主要是在正确表达原文内容（即"信"或"忠实"）的基础上，求其用词上准确无误，文字上通顺易懂，中文意思上合乎逻辑规范，技术上符合习惯说法，专业术语要正确（即"通顺"）。科技翻译在译文"优雅"或"美"的标准上，较之文学作品的翻译，要求可以稍微降低一些，主要在于忠实而准确地传达信息。因此，科技翻译的标准也可以概括为"忠于原文，表述流畅，译文地道，逻辑清晰"。

4 翻译的种类

翻译的种类可以从翻译的出发语和归宿语、翻译的手段、内容、方法、范围等各种不同的角度来予以划分。下文将对此逐个予以介绍。

首先，从出发语和归宿语的角度来看，翻译可分为外语译为本族语和本族语译为外语。如果将英语材料翻译为汉语材料，则简称为英译汉；如果将汉语材料翻译为英语材料，则简称为汉译英。依此类推，我们有法译汉、汉译法、德译汉、汉译德、日译汉、汉译日等。

从翻译的手段来看，我们有口译、笔译和机器翻译。口译（interpretation）就是指不借助于文本的口头翻译，是一种面对面的话语交流。导游、解说员等所从事的翻译工作一般为口译。笔译（translation）是指借助于文本的书面翻译。不论是口译还是笔译都是通过人来实现的。如果翻译过程通过装备有人工智能的仪器设备来实现，则被称为机器翻译。

从翻译的内容或对象来看，我们有文学翻译、政治文体翻译、科技翻译和事务性函电翻译等。文学翻译是指对文学作品的翻译。这是一种创造性的劳动，是一种有严格限制的创作。它是一种艺术，是一种难度较高的艺术工作。文学翻译不仅译原作的字句，而且要译出原作的思想、情感和精神；不仅要再现原作中的人物形象、故事情节，而且要再现原作的艺术意境，使译文读者像读原作一样得到启发和感受。政治文体翻译主要是指对政治生活中一些格式化的文体的翻译，事务性函电翻译是指对工作中的一些信函、电传、电报等的翻译，这里不再赘述。科技翻译将在下一节单独予以介绍（**参见本章第5节**）。

从翻译的方法来看，可分为直译（literal translation）和意译（free translation or paraphrase）。直译是指要保留原文的语法结构形式和原文词语的书面意义，力争译文既忠实于原文的内容又忠实于原文的形式。意译就是译文忠实于原文的内容，而不拘泥于原文的形式，翻译时可以进行词语的省略和删补，词义的转换和延伸，语序的变动，句子结构和表达方式的转换等。至于在翻译时该用哪种方法，一般的原则是，能直译的就直译，需要意译时就意译，有时两种方法只选其一，有时两种方法交替或同时使用。为了使译文趋于通顺、易读，很多时候是倾向于意译，但直译和意译不是完全独立的，它们往往彼此穿插结合，对于长句和难句的处理尤是如此。

从翻译的范围来看，可分为全文翻译（full translation）和部分翻译（partial translation）。前者是指"原语文本的每一部分都要用译语文本的材料来替代"；后者则指"原语文本的某一部分或某些部分是未翻译的，只需把它们简单地转移并掺和到译语文本中即可"。部分翻译并非"节译"，而是因为种种原因某些词不译或不可译，只能原封不动地搬进译文。

不难看出，科技英语翻译主要是从翻译的内容来划分的。本书所讲的科技英语翻译主要是指将一些用英语写成的科技方面的材料翻译成相应的汉语，从出发语和归宿语的角度，属于英译汉；从翻译的手段来看，主要谈的是笔译。在科技英语翻译中，由于科技文体的特点，较多采用直译的方法，必要的时候采用意译的方法。

5 科技翻译的特点

科技翻译的主要任务是翻译科技论文、科技报告、科技期刊文章、科技会议资料、学位论文、科技图书、专利说明书、产品说明书、操作维修说明书、标准、规格、技术档案、工艺规程、样本和图纸等科技文献。

科技文体本身包含种种变体，如科学理论专著就具有同哲学著作相似的文体特点。在阐述观点时，它有哲学文体结构严密、论述合乎逻辑的特点，而在引发问题时，它又有文学文体具有的那种生动活泼、富于想象、含蓄幽默的特点。各种变体的公用系统是语言共核，所以说科技文体由语言共核部分和共核以外的特殊部分组成。科技翻译理论既包括一般翻译理论，又包括特殊的翻译理论，即科技翻译的特点。

科技文体的基本特点是语言朴素简明，感情成分较少，多用叙述体或说明体，语法结构有规律，被动语态多，非谓语动词多，词性转换多，词义多，文章内容的专业性较强，复杂句和长句也比较多；因此，对科技文体进行翻译时要做到概念清楚，术语正确，逻辑严密，语言简练通顺，正确地表达出原文所叙述的事实和科学概念。

科技英语在描述事实时具有极为严格的客观性、准确性和严密的逻辑性，这是与普通英语的重要区别所在。而在这三性中，又尤以准确性为要。因此，科技翻译的质量标准是"信和顺"。科技翻译一定要首先做到准确。没有译文的准确性，科技翻译的意义便无可依托。译文的失之毫厘，在实际应用中就会差之千里。所谓准确性，就是译文必须忠实于原文的内容，即对原文做出准确的语际转换，力求达到与原著相同的传播效果。我国著名翻译家严复在提到译事三难中，第一难便是"信"，即准确性。由此可见他对翻译准确性的重视程度。美国语言学家奈达提出的等效翻译理论也强调翻译应是信息上而不是形式上的等同。

科技翻译不同于文学翻译。科技文章重在传达信息，因此科技翻译重在达意，要概念准确、措辞严密，但不要求语言的艺术性。文学作品重在描绘形象、抒发感情，因此文学翻译不仅要再现原作中的人物形象、故事情节，而且要保持原作的艺术境界，不仅要求语言准确严密，而且要注意辞藻文采，要求译者有更高的文字修养。从科技翻译和文学翻译的区别来看，文学翻译更多的是一门艺术，而科技翻译更多的是一门科学。艺术性地创造发挥和润饰修辞在科技翻译中较少出现。

科技翻译种类繁多，专业特点鲜明，每个专业都有其自身的技术术语。面对大量的术语，如果翻译人员对专业不熟悉，翻译时就难以得心应手。翻译一篇文章，尽管反复查阅字典，译出的文章仍很别扭，其根本原因就是对专业缺乏了解，甚至不懂。译者必须掌握一定的专业知识，透彻理解文章所叙述的专业内容，准确把握这些专业术语的内涵，才能争取用地道和精辟的汉语词汇准确地表达出原术语的概念。

外文科技文章不像文学作品有许多华丽辞藻和修饰语，它的文字表达不含蓄，语句一般比较直截了当，用词的词义也简单明白，但复杂句长句也很常见，很多句子从句多，结构复杂；因此，科技翻译需要仔细分析长句复杂句的结构，搞清句中各成分之间修饰和被修饰的关系，利用各种翻译技巧大力重组词句的顺序。

科技英语翻译的基础是良好的英语水平、较强的汉语表达能力、广博的专业知识和熟练的翻译技能。科技翻译者要掌握基本的英语句型、时态、语态等知识，还要有较高的汉语修养，能熟练地运用汉语的语法、修辞手段来准确地表达原文中各种复杂的内涵。

6 翻译的过程

翻译的过程就是在正确理解原文的基础上用另一种语言尽可能恰如其分地再现原文的过程。英译汉的过程，顾名思义，就是译者在正确理解原英文的基础上用汉语恰当地将原文传递的信息表达出来的过程。这个过程一般包括理解、表达和校改三个阶段。

在翻译实践中，这三个阶段是紧密关联、相互依存的。理解是表达的前提和基础，没有正确的理解就不可能有准确的表达；同样，没有确切的表达，正确理解的信息也不能通过翻译的过程传递给读者；校改的目的和意义就在于确保理解正确和表达恰当。理解、表达和校改还是一个往返反复和相互渗透的统一过程，译者在理解的同时也会自觉或不自觉地考虑和挑选适当的表达形式，在表达和校改的时候也会进一步加深对原文的理解。翻译的过程是一个对原文含义的理解逐步深入、对原文含义的表达逐步完善的过程。虽然这三个阶段互有渗透，但为了便于讲解，下面我们将对它们分开论述。

6·1 理解阶段

理解阶段就是运用已有的英语知识和背景知识，借助于母语对原文意思进行准确思考，脑海中显现出与英语原文相对应的汉语的意思。理解阶段是翻译过程的第一个阶段，也是最重要的阶段。对原文进行透彻的理解是确切翻译的基础和关键。多数翻译错误都是由于译者的理解失误所致。如下例。

Buffer memory read/write operations and gate opening/closing are executed at times defined by the connection to be established through the switching network.

误译：缓冲记忆读/写操作及门的打开/闭合的执行时间由开关网络需建立起的接续来规定。

这里有三处概念性误译：一是 memory，此处应是"存储器"，而不应该理解为通用英语中的"记忆"；二是 gate，此处是指"门电路"，而不应该理解为常见意义的"门"；三是 switching network，应是"交换网络"的意思，而不是"开关网络"。译者将原文中的专业词汇简单地当成普通词汇加以理解，是译文出错的根本原因。

The advent of stored program control in telephone exchanges has had as great an effect on exchange administration and maintenance as on call processing.

误译：储存程序控制的出现在电话交换机中已对交换机的管理与维护产生

巨大影响,像呼叫处理一样。

正译:在电话交换机中,储存程序控制技术的出现对交换机管理与维护的影响之大,如同它对呼叫处理一样。

上例中造成误译的主要原因是没有正确理解 have a great effect on 和 as...as... 结构的结合,而是按照字面硬译。

Energy is always leaking away and taking forms in which we can not catch it and use it. The fire under your boiler heats other things besides the water in the boiler. The steam seeps out around the piston. The air around the hot cylinder of the steam engine warms up and blows away.

原译:能量总是在不断地流失,而且它所出现的形式,往往也使我们难以捕捉和利用。你的锅炉底下的火除了加热锅炉里的水以外,还加热了一些其他的东西。蒸汽从活塞周围的缝隙渗漏出来。围绕蒸汽机灼热汽缸的空气被加热、带走。

分析:实际上,原文中的四句话是密切联系在一起的,第一句作结论,第二、三、四句话举例加以证明。译文将四句话独立开来进行分译,没有将四句话之间暗含的内在联系体现出来。

改译:能量总是以我们无法驾驭和利用的形式在不断地流失。比如,锅炉底下的火除了加热锅中的水以外,也还加热了一些其他的东西;蒸汽从活塞周围的缝隙渗漏掉;蒸汽机灼热汽缸外围的空气被加热后散发出去。

从以上三个例子我们可以看出,理解上的偏差可能出现在词语层面(例1),也可能出现在对句法(例2)和章法(例3)的理解上。因此,要正确理解原文,就必须依据上下文关系吃透原文词义,辨明原文句法语法关系,理清原文的篇章结构和语句段落之间的关系。

在词语理解方面,我们知道,大多数单词都是一词多义的,但每个单词在一定的句子中只能有一种含义。这个时候我们就要借助于上下文。所谓上下文,可以是指一个词语,一个句子,一个段落,也可以指全文或全书。如下例。

Unit Two　Science and Technology in the Past

These terminals may be local, i.e., within a few tens or hundreds of meters of the rest of the exchange, or remote.

"rest"一词原意为"休息",但在上句话中,我们应根据上下文把这个表意模糊的词用汉语里含义明确具体的词加以表达,比如说译为"停放处"。又如下例。

Laser, its creation being thought to be one of today's wonders, is nothing more than a light that differs from ordinary lights.

例句中的"laser"一词可以表示"激光",也可以表示"激光器"。将上句译为"虽然激光器的发明被认为是当代奇迹之一,其实它不过是一种光",表面上看来"激光器的发明"译文通顺,但细究起来,后半句译文成了"它(激光器)不过是一种光",不合逻辑。因此,原译文可改译为"虽然激光器的发明被认为是当代的奇迹之一,其实激光只是一种光,只不过与普通光有所不同而已"。再如下例。

If the fluid reacts with active metals, such as silver and copper alloys, pump and valve designs change a lot.

上例中的"design"一词可译为"设计""型号""形状"等,但根据上下文,我们知道,"design"应为流体和活泼金属发生反应后所"改变(change)"的对象,按照逻辑,应为泵和阀的"形状"。

除了根据上下文词语搭配和内在逻辑确定词义外,我们还应根据汉语搭配习惯选择词义,如"a **high** mountain"译为"高山",而"**high** current"则译为"**强电流**"。另外,我们还可以从专业角度、逻辑方面、语法方面来确定词义,这里不再赘述。

要做到句子层面的正确理解,我们就必须有扎实的句法知识,要准确地掌握一些固定搭配，特别是当搭配相隔较远时。我们要培养分析长难句的能力。在分析长句时，我们应首先找出原句的主干部分，然后分析原句的从属部分，弄清它们之间的修饰关系和内容大意。

　　Intel warrants that the Pentium or Pentium Pro processors, if properly used and installed, will be free from defects in material and workmanship and will substantially conform to Intel's publicly available specifications for a period of three years after the date the Pentium or Pentium Pro processor was purchased (whether purchased separately or as part of a computer system).

　　仔细分析不难发现，本句从结构上看，是由一个简短的主句和一个包含着两个省略了的条件句及一个定语从句构成。搞清了句子结构，理解和表达就方便得多。

　　说到篇章上的理解，主要是要学会在语篇层次上的理解表达。我们看一篇文章，并不是去理解一个个孤立的句子，而是要挖掘句子间的内在联系。把这些独立的句子译成汉语时，必须使用一些具有连接作用的词将这些句子连接起来，使那些暗含在上下文之间的内在联系得以充分体现出来。

6.2 表达阶段

　　表达阶段就是译者把自己从原文所理解的内容用本族语言重新表达出来。在理解阶段，译者注意力的焦点是原作者，在分析理解原作者的表达方式的基础上力图弄懂原作者想要传达的内容。在表达阶段，译者注意力的焦点是译文的读者，这一阶段应着重考虑如何用读者能自然接受的语言，告诉他们原作者所说的内容。表达的好坏取决于译者对原文理解的程度以及译者自身本族语的掌握程度。

　　表达是理解的结果，只有充分理解才有可能正确表达。如果理解发生偏差，再通顺的译文也没有意义，因为它没有达到翻译的首要标准"忠实"（和原文内容相符）。错误的理解必然导致错误的翻译。然而，正确的理解并不一定就有好的表达，这还取决于译者的汉语水平。不通顺、累赘、搭配不当、不合逻辑、歧义等都是影响译文质量的常见问题。如下例。

Unit Two　Science and Technology in the Past

Instruments capable of recording, with reasonable accuracy, the maximum pressure and muzzle velocities have been available for a number of years.

原译：仪器设备能够记录，有合理的准确，最小压力和初速一直存在好多年。

分析：译文在很大程度上歪曲了原文的意思，而且表述不够通顺流畅。

改译：能以较高的精度记录最大压力和初速的仪器已经使用多年了。

An atom consists of the nucleus and one or more electrons moving around it.

原译：一个原子由原子核和一个或更多个绕着原子核运动的电子所组成。

分析：译文表述虽然正确，但未免太啰嗦了点，在这个短句的译文中，仅"原子"一词就用了三遍，实在不够经济紧凑。

改译：原子由核和一个以上绕核运动的电子组成。

The most effective method of removing this acid contaminant is to cool and then neutralize the exhaust gases.

原译：去除这种酸性污染物的最有效方法是冷却后中和废气。

分析：译文容易使人误解为先将污染物冷却，然后对废气进行中和。

改译：去除这种酸性污染物的最有效方法是对废气进行冷却，然后加以中和。

总之，要想做到表述正确得体、通顺流畅，我们就要不断努力提高自己的汉语水平，并进行大量的翻译练习。我们的表述一定要建立在正确理解的基础之上，表达要合乎逻辑，合乎汉语的规范，合乎科技语体的特征，在说清原文内容的前提下力求做到简洁明了、通顺流畅。在翻译实践中，我们应该避免以下两个倾向：一个倾向是喜欢使用文言词语，译文中文白混杂，读起来不顺当，意思也不清楚；另一个倾向是喜欢堆砌辞藻，尤其是四言成语用得过多，往往因词害意，甚至把原意歪曲了。

6.3 校改阶段

校改阶段是理解与表达的进一步深化，是对原文内容进一步核实以及对译文语言进一步推敲的阶段。翻译时，只有理解了，表达了，才需要校改。从这个意义上说，校改是翻译工作的第三个环节。但是，把校改阶段看成是翻译的最后一个环节的观点是不客观的。我们在翻译时尽管十分细心，但译文难免会有错漏或字句欠妥的地方，校核工作能够帮助我们发现问题和不足；但要改正错误，要使译文符合忠实、通顺的标准，就需要我们进一步理解原文，推敲译文。因此，理解、表达、校改三个阶段彼此交叠，统一于翻译实践之中。

在校改阶段主要侧重两方面的工作：一是核对原文的理解；二是改换不太妥的表达。科技翻译应以准确、通顺为标准，为此，在审校英汉科技译文时，应从两个方面入手。首先，译者必须正确理解原文的语法关系、逻辑关系和词义；其次，译文不必严格区分是意译还是直译，以准确、通顺为原则。因此，我们重点要关注：

（1）原文中多义词的词义选择是否正确；

（2）原文中的内在联系是否梳理清楚，句间修饰和被修饰的关系是否理解正确；

（3）原文中的内容是否有漏译；

（4）译文是否符合汉语规范；

（5）译文中的术语是否地道；

（6）补译漏译之处，修改不妥之处，定稿。

到目前为止，我们对翻译的目的、翻译的性质、翻译的标准、翻译的种类、翻译的过程等基础知识有了一个较全面的了解。在翻译实践中，我们还有很多细节问题需要关注，还有很多原则需要遵循。比如说，我们要克服望文生义，要小心文字表层的陷阱，要把握词义理解，辨清词性结构，分清替代词语，确定句子成分，分析修饰关系，理顺句子结构等。我们还要学习各种各样的翻译方法和技巧，如科技术语的翻译、词类转译、增词法、减词法、被动语态的翻译、定语从句的翻译、状语从句的翻译、长句短译、长句分译等。这些方法将在后文予以详细介绍。

Unit Two Science and Technology in the Past

总之，翻译并不是一件很轻松的事，它需要我们打下较强的双语基础，掌握较娴熟的翻译技巧。科技翻译者必须要在具备广博知识的基础上，达到相应的专业深度。科技翻译工作者应该是通才基础上的专才，杂家基础上的专家，这就是我们努力的方向！

Quiz

1. Try to correct the mistakes in the following translation of the sentences.

    (1) Scientific discoveries and inventions do not always influence the language in proportion to their importance.

    科学的发现与发明，就其重要性的比例而言，并不一定对语言有什么影响。

    (2) The importance of superconductors in the uses of electricity cannot be overestimated.

    超导体在电器应用上的重要性不能被估计过高。

    (3) Therefore we say there is something in addition to the pressure that determines the amount of the current of electricity.

    因此我们说，除了能确定电流大小的电压外，还有一些东西。

    (4) Had telemetry been applied sooner, a number of promising aircraft developments might not have been canceled, some crashes surely would have been avoided.

    遥测技术用得较早，许多有前途的飞机研制工作未能取消，从而避免了某些坠毁事故。

    (5) The pilot lamp stopped to represent the termination of the operation.

    指示灯停止显示操作终止。

(6) If a designer were to design a bracket to support 100 lb. when it should have been figured for 1,000 lb., failure would be forthcoming.

如果设计者所设计的托架能支撑 100 磅，当它被设计为 1,000 磅时，事故一定会出现。

(7) All substances will permit the passage of some electric current, provided the potential difference is high enough.

只要有足够的电位差，所有的物体都允许一些电流通过。

(8) The heat loss can be considerably reduced by the use of firebricks round the walls of the boiler.

通过在炉壁周围使用耐火砖，热损失可以被大大降低。

(9) Television is the transmission and reception of images of moving objects by radio waves.

电视是通过无线电波的活动物体的图像的传播和接受。

(10) Afterwards he found a sunspot which lived long enough to disappear from view on the western limb of the sun, to reappear on its eastern limb, and finally to regain its old position.

后来，他发现有一个停留在太阳西部的边缘上的黑子过了很长一段时间才消失在太阳的东部边缘上，最后再次获得了原位。

Part B Reading

Science and Technology in Traditional China

Until relatively recently the general belief in the West was that the science and technology which had existed in traditional China was of relatively little importance compared with that of Europe. Now we know that this is not true. Traditional China had developed a **substantial** body of knowledge about many scientific and technological topics. Much of this knowledge **predates** that of Europe, in some cases by several centuries, and was acquired in a society which knew very little, if anything, of what was taking place in Europe.

These three discoveries—printing, gunpowder and compass—were all made much earlier in China than in Europe but, unlike in Europe, these discoveries were not followed by major changes in the structure of Chinese society.

Block printing was invented in China in the 9th century AD and printed books began to appear in the later years of that century. The oldest surviving printed book is a **Buddhist** text, which dates from AD 868, and a complete printed edition of the classical books of Confucius[1] was **commissioned** in AD 932 and completed in AD 953. Moveable type was developed in the 11th century, even though a separate piece of type was needed for each of the thousands of characters. Moveable type was not introduced in Europe until 400 years later, when Gutenberg[2] printed his Latin Bible in 1456.

The origins of gunpowder in China also date from the 9th century AD. The earliest written **formula** for a form of gunpowder, a mixture of **charcoal**, **saltpeter** and **sulphur**, appeared in a Chinese book published in AD 1044. It was not until the early 14th century that any similar reference could be found in Europe. And the new invention was soon applied to weapons such as the rocket launcher and the **barrel** gun.

The first mention of magnetism and the equivalent of a magnetic compass are even earlier. There is a reference to a "south-controlling spoon" in a text dating from AD 83. The spoon itself was carved from **lodestone** and, when placed on a highly polished **bronze** plate, always rotated until it pointed south. Chinese compasses always point south! There are many references to a "south-pointer" in the following centuries—well before the first European mention of magnetic polarity in 1180.

It seems likely that these magnetic compasses were used in **navigation** as early as the 10th century, and there is some evidence that the Chinese knew of magnetic declination—the fact that compasses do not point exactly North-South and that the difference varies with time—before Europeans knew of magnetic polarity.

New Words and Expressions

substantial 大量的，众多的	saltpeter 硝石
predate 早于	sulphur 硫磺
Buddhist 佛（教）的	barrel 枪管
commission 委托制作（交付印刷）	lodestone 天然磁石
formula 配方	bronze 青铜
charcoal 木炭	navigation 航海

Notes

1. Confucius 孔子（公元前 551—公元前 479），春秋末期思想家、政治家、教育家，儒学学派的创始人。

2. Gutenberg 谷登堡（1398—1468），德国金匠，发明活字印刷术，排印过《42 行圣经》。

Unit Two　Science and Technology in the Past

📝 Exercises

1. **Find out the English equivalents of the following Chinese terms from the passage.**

   （1）印刷术　　　　　　　　（2）火药
   （3）指南针　　　　　　　　（4）中国的社会结构
   （5）刻版印刷　　　　　　　（6）佛经
   （7）儒家经典著作　　　　　（8）活字印刷
   （9）磁力罗盘　　　　　　　（10）指南勺
   （11）磁偏角　　　　　　　 （12）磁的二极性

2. **Translate the following sentences into Chinese.**

   (1) Until relatively recently the general belief in the West was that the science and technology which had existed in traditional China was of relatively little importance compared with that of Europe.

   (2) These three discoveries—printing, gunpowder and magnetism—were all made much earlier in China than in Europe but, unlike in Europe, these discoveries were not followed by major changes in the structure of Chinese society.

   (3) Moveable type was developed in the 11th century, even though a separate piece of type was needed for each of the thousands of characters. Moveable type was not introduced in Europe until 400 years later, when Gutenberg printed his Latin Bible in 1456.

   (4) The spoon itself was carved from lodestone and, when placed on a highly polished bronze plate, always rotated until it pointed south.

   (5) It seems likely that these magnetic compasses were used in navigation as early as the 10th century, and there is some evidence that the Chinese knew of magnetic declination—the fact that compasses do not point exactly North-South and that the difference varies with time—before Europeans knew of magnetic polarity.

Part C Extended Reading

List of Chinese Inventions

China has been the source of many significant inventions, including the Four Great Inventions of Ancient China: papermaking, the compass, gunpowder, and printing (both woodblock and movable type). The list below contains these and other inventions.

The Chinese invented technologies involving mechanics, hydraulics, and mathematics applied to horology, metallurgy, astronomy, agriculture, engineering, music theory, craftsmanship, nautics, and warfare. By the Warring States Period[1] (403 BC—221 BC), they had advanced metallurgic technology, including the blast furnace and **cupola furnace**, while the **finery forge** and puddling process were known by the Han Dynasty[2] (206 BC—AD 220). A sophisticated economic system in China gave birth to inventions such as paper money during the Song Dynasty[3] (960—1279). The invention of gunpowder by the 10th century led to **an array of** inventions such as the fire **lance, land mine**, naval mine, hand **cannon**, exploding cannonballs, **multistage** rocket, and rocket bombs with **aerodynamic** wings and explosive **payloads**. With the navigational aid of the 11th-century compass and ability to steer at high sea with the 1st-century **sternpost rudder**, premodern Chinese sailors sailed as far as East Africa and Egypt. In water-powered clockworks, the premodern Chinese had used the escapement mechanism since the 8th century and the endless power-transmitting chain drive in the 11th century. They also made large mechanical **puppet** theaters driven by waterwheels and carriage wheels and wine-serving automatons driven by paddle wheel boats.

The contemporaneous Peiligang and Pengtoushan cultures[4] represent the oldest Neolithic cultures of China and were formed around 7000 BC. Some of the first inventions of Neolithic, prehistoric China include semilunar and rectangular stone knives, stone hoes and spades, the cultivation of **millet**, rice and the soybean, the refinement of sericulture, the building of **rammed** earth structures with lime-plastered house floors, the creation of the potter's wheel, the creation of pottery with cord-mat-basket designs, the creation of pottery tripods and pottery steamers, and the development of ceremonial vessels and **scapulimancy** for purposes of **divination**. Francesca Bray argues that the domestication of the ox and buffalo during the Longshan culture[5] (c. 3000 BC—c. 2000 BC) period, the absence of Longshan-era irrigation or high-yield crops, full evidence of Longshan cultivation of dry-land cereal crops which gave high yields "only when the soil was carefully cultivated", suggest that the plow was known at least by the Longshan culture period and explains the high agricultural production yields which allowed the rise of Chinese civilization during the Shang Dynasty[6] (c. 1600 BC—c. 1046 BC). With later inventions such as the multiple-tube seed drill and heavy **moldboard iron plow**, China's agricultural output could sustain a much larger population.

For the purposes of this list, inventions are regarded as technological firsts developed in China, and as such does not include foreign technologies which the Chinese acquired through contact, such as the windmill from the Middle East or the telescope from early modern Europe. It also does not include technologies developed elsewhere and later invented separately by the Chinese, such as the **odometer** and chain pump. Scientific, mathematic or natural discoveries, changes in minor concepts of design or style and artistic innovations cannot be regarded as inventions and do not appear on the list.

New Words and Expressions

cupola furnace 冲天炉，熔铁炉
finery（熟铁）精炼炉
forge 锻造，铸造
an array of 一批，一系列
lance 长矛
land mine 地雷
cannon 大炮
multistage 多节的；多阶段的
aerodynamic 空气动力学的；航空动力学的
payload 火箭所载弹头；弹头内的炸药

sternpost 船尾骨
rudder（船的）舵；（飞机的）方向舵
puppet 木偶
paddle 桨
millet 黍，稷；黍的谷粒
rammed 冲压成的；捣打成的
scapulimancy 肩胛骨占卜术
divination 预言，预见，预测
moldboard 犁板
iron plow 铁犁
odometer 里程计

Notes

1. Warring States Period（公元前 403—公元前 221）战国时期，或称战国时代，简称战国，是中国历史上东周的一段时期（秦统一中原前），这一时期各国混战不休，故被后世称之为"战国"。

2. Han Dynasty（公元前 202—公元 220）汉朝，是中国历史上继秦朝之后出现的朝代，在中国历史上极具代表性，并且扮演了承前启后的重要角色。汉朝分为西汉（公元前 202—公元 9）与东汉（公元 25—公元 220）两个历史时期，合称两汉。汉朝是中国古代史上空前强大的时期，创造了灿烂辉煌的文明，当时的中国与稍晚兴起于欧洲的罗马帝国遥相并立。

3. Song Dynasty（960—1279）宋朝，是中国历史上的一个朝代，分为北宋（960—1127）与南宋（1127—1279），合称两宋。相对而言，宋朝是中国古代历史上经济与文化教育最繁荣的时代。著名史学家陈寅恪言："华夏民族之文化，历数千载之演进，造极于赵宋之世。"

4. Peiligang and Pengtoushan cultures 裴李岗文化和彭头山文化。裴

李岗文化是中国黄河中游地区的新石器时代文化，在仰韶文化之前，由于最早在河南新郑的裴李岗村发掘并认定而得名。彭头山文化是在中国湖南及湖北境内发现的史前文化，因其位于湖南省澧县澧阳平原中部的彭头山遗址而得名。公元前76世纪至公元前62世纪，为长江文明中目前已知最早的新石器时代文化。

5. Longshan culture 龙山文化，是中国新石器时代晚期的文化之一，又名"黑陶文化"，1928年因山东济南龙山镇城子崖遗址的发掘被人们发现，分布于黄河中下游，包括河南禹州的瓦店遗址。龙山文化是中国青铜器文化的形成期。

6. Shang Dynasty（约公元前1600—约公元前1050）商朝，又称殷、殷商，是中国第一个有文字记载的王朝。关于商朝的文献资料，现在已知的多来自其后面的周朝的纪录、汉朝司马迁的《史记》，以及商朝金文和甲骨文的记载。其中，甲骨文和金文的记载是目前已经发现的中国最早的成系统的文字符号。

Exercises

1. **Translate the following sentences into Chinese.**

    (1) A sophisticated economic system in China gave birth to inventions such as paper money during the Song Dynasty (960—1279). The invention of gunpowder by the 10th century led to an array of inventions such as the fire lance, land mine, naval mine, hand cannon, exploding cannonballs, multistage rocket, and rocket bombs with aerodynamic wings and explosive payloads.

    (2) Francesca Bray argues that the domestication of the ox and buffalo during the Longshan culture (c. 3000 BC—c. 2000 BC) period, the absence of Longshan-era irrigation or high-yield crops, full evidence of Longshan cultivation of dry-land cereal crops which gave high yields "only when the soil was carefully cultivated", suggest that the plow was known at least by the Longshan culture period and

explains the high agricultural production yields which allowed the rise of Chinese civilization during the Shang Dynasty (c. 1600 BC—c. 1050 BC).

(3) For the purposes of this list, inventions are regarded as technological firsts developed in China, and as such does not include foreign technologies which the Chinese acquired through contact, such as the windmill from the Middle East or the telescope from early modern Europe.

Unit Three
Language of Science

Part A Lecture

科技术语的翻译

随着新学科、新技术、新材料、新设备、新工艺的不断产生，科技新名词、新术语大量涌现。目前使用的英汉科技词典，包括那些专业性很强的分类词典，已很难全面满足科技翻译的需要，因此，掌握一定的科技术语翻译原则和方法对于科技翻译者有着重要的现实意义。

科学和技术的发展不仅为科技英语提供了极其丰富的词语，也是现代英语新词首要的、最广泛的来源。英语科技词汇产生的数量多、速度快，这些英语科技词汇的形成和扩展有两种主要途径：一种是非科技术语转化为科技术语，即常用词汇的专业化和同一词语词义的多专业化；另一种是通过传统的构词法合成科技术语。因此，从事科技翻译的人员须了解由这两种方法形成的科技术语的翻译方法。

1 非科技术语转化为科技术语的翻译方法

科技英语大量使用科技术语，其中有相当数量的科技术语是由非科技术语转化而成的，包括常用词汇的专业化和同一词语词义的多专业化。

1·1 常用词汇的专业化

常用词汇的专业化是指英语的常用词用到某一专业科技领域中成为专业技术词汇，并具有严格的科学含义。如名词 pupil 在日常用语中为"小学生"，在解剖学上称为"瞳孔"。又如 carrier（携带者），在医学科学上意为"带菌体"，具有明显的生物学、医学色彩。

1·2 同一词语词义的多专业化

随着常用词汇的专业化，同一个英语的常用词不仅被一个专业采用，而且被许多专业采用来表示各自的专业概念,甚至在同一专业中同一个词又有许多不同词义。如名词 cell 在日常用语中为"小室",在生物学上为"细胞",在电学上为"电池",在建筑上为"隔板"。又如 power,在机械动力学这个专业中,它的词义就有"力""电""电力""电源""动力""功率"等。再如动词 feed,在一般用语中为"进食",而在科技方面可以理解为"供电""加水""添煤""上油""进刀"等。试看下面一组例句:

She *feeds* the cow with bran. 她用麸皮喂牛。

This motor can *feed* several machines. 这台电动机可以给几台机器**供电**。

We use pumps to *feed* fresh water into the boiler. 我们用抽水机给锅炉**加水**。

They use mechanical stockers to *feed* coal into the furnace. 他们用自动加煤机给炉子**添煤**。

We can *feed* the oil into the bearing in several ways. 我们可以用几种方法给轴承**上油**。

Tool carriage supports and *feeds* the cutting tools over the work. 刀架托住刀具对工件**进刀**。

1·3 翻译方法

翻译这类科技术语必须根据所涉及的专业内容确定词义,然后选用与之相当的汉语词语来翻译,否则虽然是个别词语,译法不当也会使整个句子的意思变得不好理解,甚至造成误译。所以,遇到不懂的或者不大熟悉的专业词语,一定要仔细查阅有关的词典或专业书刊,选用恰当的译法。

2 合成科技术语的翻译

英语科技术语按形态可以分为三种类型,即单词型、合成型和短语型。这三类术语的形成都是按英语构词法的基本构造规律产生的。合成科技术语的构词及翻译方法有以下几种。

2.1 合成法

合成法是指将两个或两个以上的旧词组合成一个新词。科技英语中的合成词有合写式（无连字符）与分写式（有连字符），如：

fallout 放射性尘埃（合写式）

heat-wave 热浪（分写式）

大部分科技英语合成词均采用直译法，即将两个合成语素的词义直译作偏正连缀。有些合成词汉译时需适当增词，才能准确通顺地表达原意。

2.2 混成法

混成法是指将两个词在拼写或读音上比较适合的部分以"前一词去尾、后一词去首"，加以叠合混成，混成后新词兼具两个旧词之形义，如：

smog=smoke+fog 烟雾

medicare=medical+care 医疗保健

不少混成词前一词或后一词可能是一个完整的单音节或双音节词，如：

escalift=escalator+lift 自动电梯

混成词在科技英语中似乎比在其他语类中用得普遍，原因可能是熟悉本专业知识或技术的人认为混成词比较简略而不难借助联想理解词义，如 biorhythm 是由 biological 与 rhythm 混成的，词义是"生理节奏"；mechnochemistry 是由 mechanical 与 chemistry 混成的，词义是"机械化学"等。

混成词是合成词的一种变体，翻译时注重直译，一般取偏正式复合构词，如自动（偏：修饰语）电梯（正：中心词）。

2.3 词缀法

词缀法是指利用词缀（前缀或后缀）作为词素构成新词。英语许多词缀的构词能力很强，因此词缀法就成为科技英语构词的重要手段。如前缀 anti-（反）加在根词前即构成 antimatter（反物质）、antismog（防烟雾剂）、antiparticle（反粒子）、antipollution（反污染）、antihyperon（反超子）、anticyclone（反气旋）等。

利用词缀法构词的优越性是显而易见的。词缀具有极大的灵活性，同时又具有极强、极广泛的搭配表意能力，这是因为一方面词缀的基本词义都比较稳定、明确；另一方面它们的附着力都很强，附着在根词之前或之后，概念可以立即形成。此外，拉丁语源的词缀本来就十分丰富，随着科技和英语的发展，又不断产生新的构词成分，并被广泛利用作为科技词语的构词手段，如 bio-（生命、生物）、thermo-（热）、electro-（电）、aero-（航空）、carbo-（碳）、hydro-（水、氢）、ite-（矿物）、-mania（热、狂）等。这些构词成分并不是传统的英语词缀，它们都是一些科技词汇的词头或词尾，但它们的黏附构词能力是很强的。熟悉这些科技词汇的构词成分，对解析、判断、翻译英语科技词语是很有帮助的。

许多长科技词语都可以用解析法析出词义。如：

barothermograph=baro（气压）+thermo（温度）+graph（记录器）：气压温度记录器

2·4 缩略法

缩略法是指将某一词语组合中主要的词的第一个字母组成新词的构词方法。科技英语中常用缩略词是因为它们简略、方便，典型的例子是 laser（激光，由 light amplification by stimulated emission of radiation 这个词语组合缩略而成）。常用的科技缩略词有 ADP（automatic data processing 自动数据处理）、IC（integrated circuit 集成电路）、DC（direct current 直流电）、AC（alternating current 交流电）等。值得一提的是，英译汉时如果遇到没有把握的缩略词，必须要多查几本词典。

2·5 借用专用名词

许多科技名词和术语是借用了专用名词（包括人名、地名、商名、商标、机构等）。遇到这种词也宜多查词典。由于它们是专用名词，因此一般取音译或音译加注释。xerox（静电复制）原是美国的商号名，现在已成为普通名词，并派生了 to xerox、xeroxer 等词。

除了以上几种主要的构词法以外，科技英语还利用剪截法（如从

laboratory 剪截出 lab）、逆序法（如从 laser 逆生出 to lase）、造词法（如物理学中的 quark）、词性转换（如 to contact the terminal 中的 contact 已由传统的名词词性转换成动词）以及借用外来语（如 gene，借自德语）等构词法构成通用或专用词语。

3 科技术语译名的统一

由于科技发展日新月异，科技术语大量涌现，术语使用范围和使用频率日益扩大和增加，科技术语存在一个命名的统一问题。

根据汉语习惯和翻译传统，统一译名的原则大致有译义、译音、音义兼译、译形（象译）等。

3.1 译义

译义是根据原词的含义译成相对应的汉语。这样翻译词义明显，易于理解和接受，所以在可能的情况下，应该尽量采用这种译法。如下例。

gramophone 留声机

telescope 望远镜

radio 收音机

bicycle 自行车

thermometer 温度计

helicopter 直升机

jet 喷气式飞机

3.2 译音

人名、地名一般是译音的，还有一些不便译义或者一时找不到相对应的汉语来译的词也常用音译法进行翻译。如下例。

A. 人名：

Watt 瓦特

Ohm 欧姆

Newton 牛顿
Einstein 爱因斯坦
B. 地名：
London 伦敦
Canada 加拿大
C. 计算单位：
volt 伏特（电压单位）
ampere 安培（电流单位）
joule 焦耳（功的单位）
hertz 赫兹（频率单位）
D. 其他：
morphine 吗啡
coffee 咖啡
aspirin 阿司匹林

译音的词由于词义不易理解，往往要经过广泛使用，为大家所接受，才会成为比较确定的译名，如"咖啡""尼龙""雷达""滴滴涕"等。

3·3 音义兼译

有些词是采用译音与译义相结合的方法翻译的。这种译法是在音译之后加用一个表示特点的词，或者一部分译音，一部分译义，使词义比单纯译音较易理解。如下例。

A. 译音后加用表示特点的词：
beer 啤酒
jeep 吉普车
cigar 雪茄烟
flannel 法兰绒
B. 一部分译音，一部分译义的词：
motorcycle 摩托车
neon sign 霓虹灯
Einstein equation 爱因斯坦方程
Kilowatt 千瓦

3.4 译形（象译）

英语用字母或词描述某种事物的外形，汉译时也可以通过具体形象来表达原义。它既不是意译，也不是音译。但作为一种翻译手段，可补意译和音译的不足，称为"象译"。大致有下面三类。

A. 以丁、工、人、十、之等汉字表达形象：
T-plate 丁字板
I-steel 工字钢
cross-wire 十字线
herring bone gear 人字齿轮
zigzag road 之字路

B. 直接以拉丁字母表达形象：
X-tube X 形管
U-pipe U 形管
Z-beam Z 字梁
O-ring O 形环

C. 借用各种事物的形象来表达：
U-bolt 马蹄螺栓
V-belt 三角皮带
X-type 交叉形

D. 其他译法，有些词由于上述译法不适用，就用保留原词不译的办法来处理：
X ray X 射线
Y alloy Y 合金
Z axis Z 轴

3.5 汉语对应词早已有之者，不再取新译名

汉语历史悠久，许多动植物、矿物及天文、地理名称古已有之。翻译时，须沿用汉语的对应词语。如英语的 tin plate 在汉语中叫"马口铁"，the Plough 称作"北斗星"。

Unit Three　Language of Science

3·6　沿用不规范译名

约定俗成的不规范译名，由于沿用已久，不应再按以上所述的原则重新译名。如 Greenwich 仍为"格林尼治"。

从上面的介绍可看出，专业的名词术语一般都有各专业的习惯译法，翻译时应采用统一的译名。如果译名不统一，会造成混乱，影响对专业内容的理解。例如，nylon 这个词现在通译为"尼龙"，如果根据基本词典把它译成"耐纶"，或者随便译成"乃隆"，就可能被当成另一种材料。又如 laser（激光）过去译为"莱塞"，但也有人译成"镭射"，不知者就会以为是两种东西。

遇到有几种译名的同时，不要随便根据词典选用某个旧的译名，而应查阅有关的书刊，采用通行的译法。

Quiz

1. Translate the following sentences into Chinese. While translating, pay more attention to the italicized words.

    (1) Automatic cells can *operate* without being charged for decades.

    (2) Energy will *operate* some changes under this temperature.

    (3) The computer can *operate* only according to instructions.

    (4) In a future conflict, aircraft will have to *operate* under the threat of missiles.

    (5) Storage cells can be used to *operate* automobiles.

    (6) The doctor decided to *operate* on him immediately.

    (7) Rockets *operate* in the vacuum of outer space.

    (8) The old worker teaches him how to *operate* machines.

    (9) This company *operates* only domestically.

    (10) The new motor *operates* well.

Part B Reading

Numbers and Mathematics

It is said that mathematics is the base of all other sciences, and that arithmetic, the science of numbers, is the base of mathematics. Numbers consist of whole numbers (**integers**) which are formed by the digits 0, 1, 2, 3, 4, 5, 6, 7, 8, and 9 and by combinations of them.[1] For example, 247—two hundred and forty seven—is a number formed by three digits. Parts of numbers[2] smaller than 1 are sometimes expressed in terms of fractions, but in scientific usage they are given as decimals. This is because it is easier to perform the various mathematical operations if decimals are used instead of fractions. The main operations are: to add, subtract, multiply and divide; to square, cube or raise to any power; to take a square, cube or any root and to find a ratio or proportion between pairs of numbers or a series of numbers.[3] Thus, the decimal, or ten-scale, system is used for scientific purposes throughout the world, even in countries whose national systems of weights and measurements are based upon other scales. The other scale in general use nowadays is the binary, or two-scale, in which numbers are expressed by combinations of only two digits, 0 and 1. Thus, in the binary scale, 2 is expressed as 010. 3 is given as 011. 4 is represented as 100, etc. This scale is perfectly adapted to "**on-off**" pulses of electricity, so it is widely used in electronic computers: because of its **simplicity** it is often called "the lazy school-boy's dream"!

Other branches of mathematics such as algebra and geometry are also extensively used in many sciences and even in some areas of philosophy. More specialized extensions, such as probability theory and group theory, are now applied to an increasing range of activities, from economics and the design of experiments to war and politics. Finally, a knowledge of **statistics** is required by every type of scientists for the analysis of data. Moreover, even

Unit Three   Language of Science

an elementary knowledge of this branch of mathematics is sufficient to enable the **journalist** to avoid misleading his readers, or the ordinary citizen to **detect** the attempts which are constantly made to deceive him.

New Words and Expressions

> integer 整数
> on-off 开—关（的），接通—断开（的）
> simplicity 简单
> statistics 统计学
> journalist 新闻记者
> detect 察觉

Notes

1. Numbers consist of whole numbers (integers) which are formed... 一句中的限定性定语从句对其先行词 whole numbers（integers）起解释说明的作用，因此定语从句宜拆开翻译，作为补充说明的分句。全句译为：数由整数构成，整数则由 0、1、2、3、4、5、6、7、8 和 9 这些数字及其任意组合构成。

2. part of numbers 意为"数的一部分（a part of numbers）"；parts of numbers 意为"数的全部，数的各部分"，parts of numbers smaller than 1 意为"小于 1 的数"。

3. ...to find a ratio or proportion between pairs of numbers or a series of numbers 译为：计算每两个数或每组数之比率或比例。

Exercises

1. Find out the English equivalents of the following Chinese terms from the passage.

   （1）整数   （2）分数
   （3）小数   （4）平方

（5）立方　　　　　　　（6）任意次幂计算
（7）开平方　　　　　　（8）开立方
（9）求任意次方根　　　（10）十进制
（11）二进制　　　　　　（12）几何
（13）代数　　　　　　　（14）概率论
（15）群论

2. Translate the following sentences into Chinese.

(1) It is said that mathematics is the base of all other sciences, and that arithmetic, the science of numbers, is the base of mathematics.

(2) The other scale in general use nowadays is the binary, or two-scale, in which numbers are expressed by combinations of only two digits, 0 and 1.

(3) This scale is perfectly adapted to "on-off" pulses of electricity, so it is widely used in electronic computers: because of its simplicity it is often called "the lazy school-boy's dream"!

(4) Other branches of mathematics such as algebra and geometry are also extensively used in many sciences and even in some areas of philosophy. More specialized extensions, such as probability theory and group theory, are now applied to an increasing range of activities, from economics and the design of experiments to war and politics.

(5) Moreover, even an elementary knowledge of this branch of mathematics is sufficient to enable the journalist to avoid misleading his readers, or the ordinary citizen to detect the attempts which are constantly made to deceive him.

Part C Extended Reading

Mathematical Methods in Physics and Engineering

Students of physics and engineering, after having completed the standard course in calculus through differential equations, are faced with the problem of deciding what additional mathematics to take. Some **leave** mathematics **out of** their curriculum altogether, only to discover, after they have **embarked on** a program of graduate study, that they must go back and take undergraduate courses in order to acquire the mathematical background and **maturity** necessary to understand work being done in modern physics and engineering. Others, in their haste to pick up additional techniques not covered in the elementary calculus course, take courses in advanced engineering mathematics which cover such topics as Fourier series, Laplace transforms, partial differential equations, boundary-value problems, and complex variables. These topics are often presented in a very **heuristic fashion**, because the students lack a solid background in analysis. In fact, many of these mathematical techniques are taught in the same heuristic manner in courses in physics and engineering when they are needed in specific applications. Eventually, however, most graduate students will need a thorough understanding of applied mathematics if they are going to be able to read the literature in their own field and use mathematics effectively in their own work.

It is my opinion that a student who intends to do graduate work in physics or in some branches of engineering should develop a broad mathematical base for his graduate studies when he is an undergraduate. He should do this by taking a course in what has traditionally been called advanced calculus, followed by **an introduction to** mathematical physics based on the advanced calculus. Without the burden of presenting applications,[1] advanced calculus can be a course which carefully develops the concepts of function, limit, continuity, differentiation, integration, infinite series, improper integrals, and possibly functions of a complex variable. Then, with this as a background,² the applications can be presented in the mathematical physics course, and the students will be capable of understanding thoroughly the mathematics involved. This book has been written to **fill the need for** a textbook for the latter course³.

The student is assumed to have had elementary calculus through differential equations, some elementary mechanics, plus the kind of advanced calculus course described above, including vector analysis. Some knowledge of functions of a complex variable is very helpful but not essential. **The bulk of** material can be understood without prior knowledge of complex variable analysis. The necessary higher algebra is developed from the beginning, so that **no more than** elementary college algebra is needed.

New Words and Expressions

leave...out of 从……中舍去……
embark on 开始
maturity 成熟；完备，完成
heuristic fashion 启发式

an introduction to ……的入门
fill the need for 满足对……的需要
the bulk of 大部分
no more than 仅仅，不过是

Notes

1. without the burden of presenting applications 是介词短语，用作条件状语。
2. with this as a background 是一个"with+ *n.* +介词短语"的形式，句中用作方式状语。
3. 句中 the latter course 指 the mathematical physics course。

Unit Three Language of Science

📝 Exercises

1. Find out the English equivalents of the following Chinese terms from the passage.

   （1）微分方程　　　　　　　（2）傅里叶级数
   （3）拉普拉斯变换　　　　　（4）偏微分方程
   （5）边界值问题　　　　　　（6）复变函数
   （7）函数　　　　　　　　　（8）极限
   （9）连续性　　　　　　　　（10）微分
   （11）积分　　　　　　　　　（12）无穷级数
   （13）广义积分　　　　　　　（14）矢量分析
   （15）课程　　　　　　　　　（16）研究生学业
   （17）大学本科课程　　　　　（18）文献资料

2. Translate the following sentences into Chinese.

   (1) Students of physics and engineering, after having completed the standard course in calculus through differential equations, are faced with the problem of deciding what additional mathematics to take.

   (2) Eventually, however, most graduate students will need a thorough understanding of applied mathematics if they are going to be able to read the literature in their own field and use mathematics effectively in their own work.

   (3) It is my opinion that a student who intends to do graduate work in physics or in branches of engineering should develop a broad mathematical base for his graduate studies when he is an undergraduate. He should do this by taking a course in what has traditionally been called advanced calculus, followed by an introduction to mathematical physics based on the advanced calculus.

   (4) Then, with this as a background, the applications can be presented in the mathematical physics course, and the students will be capable of understanding thoroughly the mathematics involved.

Unit Four
Physical Science

Part A Lecture

词类转换法

由于英语和汉语在词法和句法上的差异,翻译中比较忌讳字对字翻译,因此翻译过程中进行词类转换是常见的。所谓词类转换,是指在翻译过程中根据译文语言的习惯进行词性转换,使译文通顺自然。下文将介绍科技英语文章翻译为汉语时常见的词类转换。

1 名词化和名词的转译

1·1 名词化

大量使用名词化结构是科技英语的特点之一。名词化就是把动词或动词短语转换为含有动作含义的名词或是名词短语。如下例。

to translate Chinese into English → the translation of Chinese into English

科技文体要求行文简洁,表达客观,信息量大,强调存在的事实,而不是某一个或一系列的动作。使用名词化结构不仅将原来的动作转化成了事实,也精简了句子的结构。变成名词短语以后,一个句子还可以容纳几个名词短语,这样可以更好地表达较为复杂的过程和思想。如下例。

例②

The ancients regarded natural processes as manifestations (表现形式) *of power by irresponsible gods*; today we think of them as manifestations of energy acting on or through matter.

Unit Four　Physical Science

 转译为动词

翻译中的词性转换存在着一定的规律，最明显的是英语中名词和介词用得多，而相比之下汉语中的动词用得多一些。因此，英译汉时，英语中的名词被转换成汉语中的动词；而汉译英时，我们往往将汉语中的动词转换成英语中的名词。

A *reduction* in condensation is achieved by the *use* of steam jackets.
使用蒸汽夹套可减少冷凝作用。

上句中的名词 reduction 和名词短语 the use of 中的 use 译为动词。此句当然也可译为：通过使用蒸汽夹套可以达到冷凝作用的减少。但这样的译文不符合中文的文法，读起来别扭，因而不宜采用。

Rockets have found *application for the exploration* of the universes.
火箭已经用来探索宇宙。

例5

Laser is one of the most sensational developments in recent years, because of its *applicability* to many fields of science and its *adaptability* to practical uses.
激光是近年来最轰动的科学成就之一，因为它可以应用于许多科学领域，也适合各种实践用途。

英译汉时，除名词转换成动词外，动名词转译为动词的情况更是比比皆是。

Reversing the direction of the current reverses the directions of its lines of force.
倒转电流的方向也就倒转了它的磁力线的方向。

The improvement of this kind of devices requires **solving** many complicated engineering problems.

改进这种装置需要**解决**许多复杂的工程问题。

当名词与其他词连用时，也可转译为相应的动词。

This oil tank **has a capacity of** ten gallons.

这个油桶可**装** 10 加仑油。

He **had a good knowledge of** chemistry.

他**精通**化学。

With the use of the increased temperature and pressure, the oil is cracked into lighter or heavier fractions.

利用高温高压，可将石油裂化成轻馏分和重馏分。

1.3 转译为形容词

在英译汉的过程中，英文句中某些做表语的名词可转换成形容词。如下例。

This experiment is an absolute **necessity** in determining the best processing route.

对确定最佳工艺流程而言，这次实验是绝对**必需的**。

The maiden voyage of the newly-built steamship was **a success**.

那艘新造轮船的首航是**成功的**。

Unit Four Physical Science

It should be noted that, because the explosives are rapidly consumed, momentary imbalances in the effective force and torque occurring during consumption are of *no significance*.

应当指出，由于炸药的迅速燃尽，所以在燃烧过程中出现作用力和力矩的瞬间不平衡是**无关紧要的**。

Without the gas pipe line, the movement of large volumes of gases over great distances would be an economic *impossibility*.

如果没有煤气管道，那么要远距离输送大量煤气从经济角度来看是**不可行的**。

1·4 转译为副词

在翻译的过程中，有时为了表达的顺畅，可将英文中的名词译为汉语中的副词。

Most of the metal ions produced during leaching are combined with sulfate as metal complexes or precipitates.

浸出过程中产生的金属离子**大都**和硫酸盐化合成金属络合物或沉淀物。

The discovery of rich petroleum resources in China is also inseparable from the oil-workers' *efforts* to "win honors".

中国丰富的石油资源的发现也是和石油工人的**努力**"争气"分不开的。

2 动词的转译

英译汉的过程中，在大多数情况下，英文中的动词有对应的汉语动词。同时，有些英文动词，尤其是一些状态动词，所包含的概念难以用汉语的动词直接表达。这时，这类动词通常转译为汉语的名词，以适应汉语的习惯表达。

Such materials are ***characterized*** by good insulation and high resistance to wear.

这些材料的**特点**是：绝缘性好，耐磨性强。

An electric current ***varies*** directly as the electromotive force and inversely as the resistance.

电流的**变化**与电动势成正比，与电阻成反比。

An electron or an atom ***behaves*** in some ways as though it were a group of waves.

电子和原子的**表现**，多少有点像一组波。

The new liquid crystals ***feature*** wide working temperature range, low voltage operation and high reliability.

新型液晶的**特点**是：工作温度范围宽，工作电压低，可靠性高。

If a box ***weighs*** 10 kilograms, an upward force of 10 kilograms must be exerted to lift it.

如果箱子的**重量**是 10 公斤，就必须施加 10 公斤向上的力才能把它举起来。

Coating thickness ***ranges*** from one tenth mm to 2 mm.

涂层的厚度**范围**是 0.1 毫米至 2 毫米。

Unit Four Physical Science

例23

If we connect one ammeter to measure the electric current flowing into a torch bulb and another one to measure the current flowing out, they each *read* the same.

如果我们连接一个安培计来测量流入手电筒灯珠里的电流,并连接另一个安培计去测量流出的电流,它们的**读数**是相同的。

例24

The circulatory system *functions* to convey material from one part of the body to another.

循环系统的**功能**就是把物质输送到身体各处。

例25

Boiling point is *defined* as the temperature at which the vapor pressure is equal to the atmospheric pressure.

沸点的**定义**就是气压等于大气压时的温度。

3 形容词的转译

在科技文体英译汉的过程中,有些形容词根据上下文的不同,同样可以转译为汉语的其他词类,最常见的是转译为名词、动词和副词。

3·1 转译为名词

英文中用作表语并且表示事物性状的形容词常可转译为汉语中的名词,使译文更为通顺且符合中文的表达。如下例。

The nature of the organism causing measles remained *unclear* until 1911, when it was proved to be a virus.

引起麻疹的有机体的性质曾一直是一个**不解之谜**,直到1911年才证实它是一种病毒。

This work is not more *elastic* than that one.
这份工作没有那份工作有**弹性**。

Glass is much more *soluble* than quartz.
玻璃的**可溶性**比石英大得多。

Some people think tomorrow's computers will be *intelligent*.
有人认为未来的计算机将具有**智能**。

The fresh nectar is very *fluid*.
新鲜花蜜的**流动性**很强。

Evidence indicates that the outer core is about twice as *dense* as the material in the mantle.
有证据表明,外地核的**密度**约为地幔成分的两倍。

TV is *different* from radio in that it sends and receives a picture.
电视和收音机的**区别**就在于电视发送和接收的是图像。

In fission processes the fission fragments are very *radioactive*.
在裂变过程中,裂变碎片具有强烈的**放射性**。

对于那些带有定冠词的形容词,一般也转译为名词。

Unit Four　Physical Science

Both the compounds are acids, *the former* is strong, *the latter* weak.
这两种化合物都是酸，**前者**是强酸，**后者**是弱酸。

X is used in mathematics for *the unknown* in equation.
在数学上，X 用来指代方程中的**未知数**。

3·2　转译为动词

在英文中，有些形容词由动词派生出，并与介词搭配使用，在这种情况下，这些形容词往往需要转译为汉语的动词。如下例。

Natural gas is often *present* in the reservoir rock.
储油岩中常**有**天然气。

If extremely low-cost power were ever to become *available* from large nuclear power plants, electrolytic hydrogen would become competitive.
如果能从大型核电站**获得**成本极低的电力，电解氢的竞争力就会增强。

The amount of Vitamin D in the milk is not *adequate* to the demand of the infant.
牛奶中维生素 D 的含量不能**满足**婴儿的需要。

Water is a substance *suitable* for preparation of hydrogen and oxygen.
水是**适于**制取氢和氧的物质。

Both of the substances are not *soluble* in water.

这两种物质并非都**溶**于水。

3·3 转译为副词

当英文中原名词已经转译为相应汉语的动词和形容词时，那么原来在英文句中修饰名词的形容词也就相应地转译为汉语的副词。如下例。

Below 4℃, water is in *continuous* expansion instead of *continuous* contraction.

水在4℃以下就会**不断**膨胀，而不是**不断**收缩。

All of this proves that we must have a *profound* study of properties of metals.

所有的一切证明我们必须**深入**研究金属的特性。

These new developments offer *dramatic* increases in the lethality of existing gun.

这些新的研究成果**大大**提高了现役火炮的作战能力。

The cause of this phenomenon was a *complete* mystery to early man.

这一现象的起因对前人来说**完全**是个谜。

4 副词的转译

在英译汉的过程中，英文中的副词也常可转译为汉语的形容词、动词和名词。

Unit Four Physical Science

4·1 转译为形容词

当英文中的动词和形容词要转译为汉语的名词时,那么修饰它们的副词就相应地转译为形容词。如下例。

This waveguide tube is **chiefly** characterized by its simplicity of structure.

这种波导管的**主要**特点是结构简单。

The magma within the earth may be **heavily** charged with gases and steam.

地球内的岩浆可能含有**大量的**气体和蒸汽。

Gases and liquids are **perfectly** elastic.

气体和液体都是**完全的**弹性体。

These modern installations affect **tremendously** the entire communication system.

这些现代化的装置对整个通信系统有**极大的**影响。

Earthquakes are **closely** related to faulting.

地震与断裂运动有**密切的**关系。

在英文中,表示地点的副词作名词的后置定语,在英译汉中,往往转译为形容词。

The temperatures and pressures required are shown in the table **above**.

所需的温度和压力已在**上面的**表格中列出。

The power station *there* supplies the electric power to the whole city.

那里的发电厂给全市供电。

The balloon will continue to go up until the gas *inside* equals that *outside* in density.

气球会上升到其**内外**的气体密度相同为止。

The equations *below* are derived from those *above*.

下面的方程式是由**上面**的那些方程式推导出来的。

The most nearby source of water for irrigation may be the ground water reservoir *underneath*.

灌溉用的最近的水源，可能就是**地面底下**的地下水水库。

4·2 转译为名词

有些副词，在原英文句子中做次要的成分，但其表示的意义和概念在整句中有重要的作用。在英译汉的过程中，往往将此类副词转译为汉语中的名词，使译文顺畅和上口。如下例。

There seem to be no other competitive techniques which can measure range as *well* or as *rapidly* as a laser can.

就测距的**精度**和**速度**而论，似乎还没有其他的技术可与激光相比。

The blueprint must be *dimensionally* and *proportionally* correct.

蓝图的**尺寸**和**比例**必须正确。

Unit Four Physical Science

Internally the earth consists of two parts, a core and a mantle.
地球的**内部**由两部分组成：地核和地幔。

Chlorine is very active *chemically*.
氯的**化学特性**很活泼。

Gold is not *essentially* changed by man's treatment of it.
黄金的**性质**并不因人们的加工而改变。

Certainly people in enterprises must be *mathematically* informed if they are to make wise decision.
当然，企业管理人员要想做出明智的决策，就必须懂得**数学**。

The frequency range of 100 to 3,200 Hz is *acoustically* important in building.
100 赫兹到 3,200 赫兹这一频率范围，在建筑**声学**中很重要。

It was not until early 40's that chemists began to use the technique *analytically*.
直到 40 年代初，化学家才开始将这种技术用于**分析**方面。

4·3 转译为动词

副词 on、off、over、up、in、out、around、behind、forward、through 等，在英文句子中用作表语或是宾语补语，往往具有动作的意义，在一定情况下，在译文中可译成汉语的动词。如下例。

An exhibition of new products is *on* in Shanghai.

新产品展示会正在上海**举行**。

The chemical experiment is *over*.

化学试验已**结束**。

When the switch is *off*, the circuit is open and electricity doesn't go through.

当开关**断开**时,电路就形成开路,电流不能通过。

Open the valve to let air *in*.

打开阀门,让空气**进入**。

In this case the temperature in the furnace is *up*.

在这种情况下,炉温就**升高**。

In a gas the molecules are relatively far *apart*.

相对来说,气体分子**离**得很远。

5 介词的转译

英文中的介词在很多情况下可转译为汉语的动词。

例69

A force is needed to move an object *against* inertia.

为使物体**克服**惯性而运动,就需要一个力。

Unit Four Physical Science

The letter E is commonly used *for* electromotive force.
字母 E 通常被用来**指代**电动势。

With its many characteristic properties iron has come to be the leading material for general engineering structures.
铁具有多种性能，所以它已成为一般工程结构中的主要材料。

Noise figure is minimized *by* a parametric amplifier.
采用参量放大器可将噪音系数降至最低限度。

以上讨论的是在科技翻译的英译汉中常见的几种词类转译的方法。词类转译在翻译中非常普遍。同时应指出的是，翻译本无定法，并不是每一个英文句子在翻译为汉语的过程中，都要符合以上论述的译法。英语翻译成汉语，就要符合汉语的语法和句法，使译文读起来没有从他国文字对译过来的痕迹。因此，一个好的译者不是生搬硬套这些从翻译实践中总结出来的理论方法，而是在学习了这些理论后，以它们为依托，积极地进行翻译实践。

Quiz

1. Translate the following sentences into Chinese. While translating, pay more attention to the italicized words.

    (1) This steam engine is only about 15 percent *efficient*.

    (2) If a body weighs 50 kilograms, it means that the earth *attracts* the body with a force of 50 kilograms.

    (3) The wave beam can be efficiently radiated from a *convenient* short antenna.

    (4) The scale under the pointer *reads* directly in amperes.

(5) A highly developed physical science is **characterized** by an extensive use of mathematics.

(6) Geology is concerned with **systematic study** of rocks and minerals.

(7) The buoyant force **up** equals the weight **down**.

(8) A **continuous increase** in the temperature of a gas confined in a container will lead to a **continuous increase** in the internal pressure within the gas.

(9) We found **difficulty** in solving this complicated problem.

(10) Optical fiber systems that can carry more telephone conversations than wire pairs are **economically** attractive.

Part B Reading

Physics

Physics is often defined as the science of matter and energy. Physics is concerned chiefly with the laws and properties of the material universe. These are studied in the closely related sciences of mechanics, heat, sound, electricity, light, and atomic and **nuclear structure**. The principles studied in these fields have been applied in numerous combinations to build our mechanical age. Such recent terms as chemical physics and biophysics **are indicative of** the widening application of the principles of physics, even in the study of living organisms.[1]

Mechanics is the oldest and basic branch of physics. This portion of the subject deals with such ideas as inertia, motion, force, and energy. Of especial interest are the laws dealing with the effects of forces upon the form and motion of objects, since these principles apply to all devices and structures such as machines, buildings, and bridges.[2] Mechanics includes the properties and laws of both solids and fluids.

The subject of heat includes the principles of temperature measurement, the effects of temperature on the properties of materials, heat flow, and thermodynamics—the study of transformations involving heat and work. These studies have led to increased efficiency of power production, the development of high-temperature alloys and **ceramics**, the production of temperatures near absolute zero, and to important theories about the behavior of matter and radiation.[3]

The study of sound is of importance not only in music and speech but also in communications and industry. The **acoustical** and communications engineer is concerned with the generation, transmission, and absorption of sound.[4] An understanding of scientific principles in sound is of importance to the radio engineer. The industrial engineer is greatly concerned with the effects of sound in producing **fatigue** in production **personnel**.

Electricity and magnetism are fields of physics which are of peculiar importance in the rapid development of technology in power distribution, lighting, communications, and the many electronic devices which provide conveniences, entertainment, and tools for investigation in other fields. An understanding of the sources, effects, measurements, and uses of electricity and magnetism is valuable to the worker in that it enables him to use more effectively the **manifold** electrical devices now so vital to our efficiency and comfort.[5]

Optics is the portion of physics that includes the study of the nature and **propagation** of light, the laws of **reflection**, and the bending or **refraction** that occurs in the transmission of light through **prisms** and **lenses**. Of importance also are the separation of white light into its **constituent** colors, the nature and types of **spectra**, **interference**, **diffraction**, and **polarization** phenomena.[6] **Photometry** involves the measurement of **luminous intensities** of light sources and of the **illumination** of surfaces.

A fascinating portion of physics is known as modern physics. This includes electronics, **atomic and nuclear phenomena**, **photoelectricity**, X-rays, **radioactivity**, the **transmutations** of matter and energy, relativity, and the phenomena associated with **electron tubes** and the electric waves of modern radio. The breaking up of atoms now provides a practical source of energy. Many of the devices that are **commonplace** today are applications of one or more of these branches of modern physics. Radio, long-distance **telephony**, **sound amplification**, and television are a few of the many developments made possible by the use of electron tubes. Photoelectricity makes possible television, transmission of pictures by wire or radio, sound motion pictures, and many devices for the control of machinery.[7] Examination of **welds** and **castings** by X-rays to locate hidden flaws is standard procedure in many industries. The practical application of the developments of physics continues at an ever increasing rate.

Unit Four　Physical Science

New Words and Expressions

nuclear structure 核结构
be indicative of 表现出……
ceramics 陶瓷
acoustical 声学的
fatigue 疲劳
personnel（全体）人员
manifold 多样的，各种各样的
propagation 传播
reflection 反射
refraction 折射现象
prism 棱镜
lens 透镜
constituent 构成的
spectra 光谱
interference 干涉（现象）
diffraction 绕射，衍射（现象）
polarization 极化；偏振（现象）
photometry 光度学，测光学
luminous intensity（发）光强（度）
illumination 照明，照（明）度
atomic and nuclear phenomena 原子和核现象
photoelectricity 光电学，光电现象
radioactivity 放射学
transmutation 转变，转换
electron tube 电子管
commonplace 平凡的，平常的
telephony 电话（学）
sound amplification 扩音器
weld 焊接
casting 铸件

Notes

1. 句中 chemical physics and biophysics 作为 recent terms 的例子，译为：像化学物理学和生物物理学这类新术语。

2. 句中主句 Of especial interest are the laws 是倒装结构，起强调作用，因此，翻译时仍保持倒装结构，译为：具有特殊意义的是……；句中 dealing with the effects of forces upon the form and motion of objects 做定语，修饰 the laws，译为：涉及力对物体的形状和运动的效应的那些定律。

3. 句中谓语 have led to 后接四个并列宾语，翻译时应保留并列结构。句中 the behavior of matter and radiation 意为"物质的性质和辐射的性质"。

4. 句中 the acoustical and communications engineer 的 the 是限定 engineer 的，意为"声学与通信工程人员"。

5. 句中 in that 为词组，意为"在于，因为"。

6. 本句为倒装句，句中主语分别是 the separation of white light into its constituent colors, the nature and types of spectra 和 interference, diffraction, and polarization phenomena。全句译为：其他重要内容还有白色光分解成单色光、光谱的本质及类型、干涉现象和衍射现象以及偏振现象等。

7. 句中宾语 television, transmission of pictures by wire or radio, sound motion pictures, and many devices for the control of machinery 太长，故将宾语补足语 possible 提到宾语前面。句型可译为：光电技术使……成为可能。

Exercises

1. Find out the English equivalents of the following Chinese terms from the passage.

   （1）力学　　　　　　　　（2）热学
   （3）声学　　　　　　　　（4）电学
   （5）热力学　　　　　　　（6）磁学
   （7）光学　　　　　　　　（8）电子学
   （9）化学物理学　　　　　（10）生物物理学
   （11）惯性　　　　　　　（12）固体
   （13）流体　　　　　　　（14）热流
   （15）X 射线　　　　　　（16）相对论

2. Translate the following sentences into Chinese.

   (1) Physics is often defined as the science of matter and energy. Physics is concerned chiefly with the laws and properties of the material universe.

   (2) The principles studied in these fields have been applied in numerous

combinations to build our mechanical age. Such recent terms as chemical physics and biophysics are indicative of the widening application of the principles of physics, even in the study of living organisms.

(3) Of especial interest are the laws dealing with the effects of forces upon the form and motion of objects, since these principles apply to all devices and structures such as machines, buildings, and bridges.

(4) These studies have led to increased efficiency of power production, the development of high-temperature alloys and ceramics, the production of temperatures near absolute zero, and to important theories about the behavior of matter and radiation.

(5) An understanding of the sources, effects, measurements, and uses of electricity and magnetism is valuable to the worker in that it enables him to use more effectively the manifold electrical devices now so vital to our efficiency and comfort.

(6) Photoelectricity makes possible television, transmission of pictures by wire or radio, sound motion pictures, and many devices for the control of machinery.

Part C Extended Reading

Relationship of Chemistry to Other Sciences and Industry

Chemistry may be broadly **classified** into two main branches: organic chemistry and inorganic chemistry. Organic chemistry is concerned with **compounds containing the element carbon.** The term organic was originally derived from the chemistry of **living organisms:** plants and animals. Inorganic chemistry deals with all the other elements as well as with some carbon compounds. Substances classified as inorganic are derived mainly from mineral sources rather than from animal or vegetable sources.

Other subdivisions of chemistry, such as analytical chemistry, physical chemistry, biochemistry, electrochemistry, geochemistry, and radiochemistry, may be considered specialized fields of, or **auxiliary** fields to, the two main branches. Chemical engineering is the branch of engineering that deals with the development, design, and operation of chemical processes. A chemical engineer usually begins with a chemist's laboratory-scale process and develops it into an industrial-scale operation.

Besides being a science in its own right, chemistry is the servant of other sciences and industry. Chemical principles contribute to the study of physics, biology, agriculture, engineering, medicine, space research, **oceanography,** and many other sciences. Chemistry and physics are **overlapping** sciences, since both are based on the properties and behavior of matter. Biological processes are chemical in nature. The **metabolism** of food to provide energy to living organisms is a chemical process. Knowledge of molecular structure of **proteins**, **hormones, enzymes,** and the **nucleic acids** is assisting biologists in their investigations of the **composition**, development, and reproduction of living cells.

Chemistry is playing an important role in **alleviating** the growing shortage of food in the world. Agricultural production has been increased with the use of chemical fertilizers, **pesticides,** and improved varieties of seeds. Chemical **refrigerants** make possible the frozen food industry, which preserves large amounts of food that might otherwise spoil. Chemistry is also producing synthetic nutrients, but much remains to be done as the world population increases relative to the land available for cultivation. Expanding energy needs have brought about difficult environmental problems in the form of air and water pollution. Chemists and other scientists are working diligently to alleviate these problems.

Advances in medicine and chemotherapy, through the development of new drugs, have contributed to prolonged life and the relief of human suffering. More than 90% of the drugs and pharmaceuticals being used in the United States today have been developed commercially within the past 45 years. The plastics and **polymer** industry, unknown 60 years ago, has revolutionized the **packaging and textile industries** and is producing durable and useful construction materials. Energy derived from chemical processes is used for heating, lighting, and transportation. Virtually every industry is dependent on chemicals—for example, the **petroleum**, steel, rubber, **pharmaceutical**, electronic, transportation, **cosmetic**, **garment**, aircraft, and television industries—and the list could go on and on.

New Words and Expressions

classify 分类
compound containing the element carbon 含有碳元素的化合物
living organism 生物体
auxiliary 辅助的
oceanography 海洋学
overlap 重叠
metabolism 新陈代谢
protein 蛋白质
hormone 荷尔蒙
enzyme 酶
nucleic acid 核酸
composition 合成物
alleviate 减轻，减缓
pesticide 杀虫剂
refrigerant 制冷剂
polymer 聚合体
packaging and textile industry 包装业和纺织业
petroleum 石油
pharmaceutical 药品，成药
cosmetic 化妆品
garment 服装

📝 **Exercises**

1. **Find out the English equivalents of the following Chinese terms from the passage.**

   （1）有机化学　　　　　　　（2）无机化学
   （3）分析化学　　　　　　　（4）物理化学
   （5）生物化学　　　　　　　（6）电化学
   （7）地球化学　　　　　　　（8）放射能化学
   （9）合成营养　　　　　　　（10）化学疗法

2. **Translate the following sentences into Chinese.**

   (1) Substances classified as inorganic are derived mainly from mineral sources rather than from animal or vegetable sources.

   (2) Besides being a science in its own right, chemistry is the servant of other sciences and industry.

   (3) Chemical refrigerants make possible the frozen food industry, which preserves large amounts of food that might otherwise spoil. Chemistry is also producing synthetic nutrients, but much remains to be done as the world population increases relative to the land available for cultivation.

   (4) Advances in medicine and chemotherapy, through the development of new drugs, have contributed to prolonged life and the relief of human suffering.

Unit Five

Earth Science

Part A　Lecture

增词法与减词法

英汉两种语言由于表达方式不尽相同，翻译时既可能要将词类加以转换，又可能要在词量上进行增减。本单元将介绍翻译过程中的增词和减词方法。

1 增词法

增词法就是在翻译时根据意义上（或修辞上）和句法上的需要增加一些词来更忠实通顺地表达原文的思想内容。增词当然不是无中生有地随意增加，而是增加原文中虽无其词但有其意的一些词，从而使得译文在语法、语言形式上符合译文习惯，并在文化背景、语言联想方面与原文一致，最终使译文与原文在内容和形式等方面对等。

1·1 根据句子结构的需要增词

翻译时，有时须根据句子结构的需要增补一些词，使句子符合汉语的表达习惯。

Were there no gravity, there would be no air around the earth.
假如没有重力，地球周围就没有空气。

The sun warms the earth, which makes it possible for plants to grow.
太阳温暖地球，从而使植物生长。

Diskette that contains a computer virus will spread the virus to the computer. The virus will infect any other diskettes placed in that computer later. Experts say that you should keep your information diskettes write-protected if you can.

带有计算机病毒的软盘会使病毒传播到计算机上。而病毒会传到其他任何使用这台计算机的软盘上。**因此**专家建议用户最好给自己的信息软盘加以写入保护。

1·2 增加概括性的词语

英语和汉语都有概括词。英语中的 in short、and so on、etc. 等，翻译时可以分别译为"总之""等等""……"，但有时英语句子中并没有概括词，而翻译时却往往可增加"两人""双方""等""等等""凡此种种"等概括词，同时省略英语中的连接词。

The frequency, wave length, and speed of sound are closely related.
声音的频率、波长与速度**三者**密切相关。

The chief effects of electric currents are *the magnetic, heating and chemical effects*.
电流的主要效应有磁效应、热效应和化学效应**三种**。

The factors, voltages, current and resistance, are related to each other.
电压、电流和电阻**这三个因素**是相互关联的。

Human beings have not yet been able to make an element by combining *protons, neutrons and electrons*.
人类还不能通过结合质子、中子、电子**三者**来合成元素。

The principal functions that may be performed by vacuum tubes are *rectification, amplification, oscillation, modulation and detection*.

真空管的**五大功能**是：整流、放大、振荡、调制和检波。

The units of *"ampere" "ohm" and "volt"* are named respectively after three scientists.

安培、欧姆、伏特**这三个单位**是分别根据三位科学家的姓氏来命名的。

1·3 根据名词复数概念增词

翻译时，有时为了明确原文的含义，需要通过增译"们""一些""许多"等词语把英语中表示名词复数的概念译出。如下例。

Air is a mixture of *gases*.

空气是**多种气体**的混合物。

Things in the universe are changing all the time.

宇宙中**万物**总是在不断变化的。

The *substances* get into the soil, into plants and into human bodies.

这些物质进入土壤、植物和人体。

In the 1980s, *departments* bought their own minicomputers and *managers* bought their PCs.

20世纪80年代，**许多部门**都购买了自己的微机，**经理们**也购买了个人电脑。

Unit Five　Earth Science

Note that *the words* "velocity" and "speed" require explanation.

请注意,"速度"和"速率"**这两个词**需要解释。

The light from a laser is like a pencil of light in that it has nearly parallel *sides*.

激光像一束**两边**都平行的锥光线。

The first electronic *computers* used vacuum tubes and other components, and this made the equipment very large and bulky.

第一**批**电子计算机使用真空管和其他元件,这使得设备又大又笨重。

However, in spite of all these similarities between a voltmeter and an ammeter there are also important *differences*.

可是,尽管伏特表和安培表之间有这些类似之处,但还有**若干重要的**差别。

1·4　增加原文中省略的词语

英语句子的某些成分如果已在前面出现,后面往往省略,但在英译汉时,一般需要将其补出。如下例。

例18

High voltage is necessary for long transmission line while low voltage for safe use.

(=...while low voltage is necessary for safe use.)

远距离输电需要高压,安全用电**需要**低压。

Plastic bowls marked microwavable are probably safer than those that aren't.

(=...safer than those plastic bowls are not marked microwavable.)

使用贴有"可用于微波炉烹调"标志的塑料碗也许比使用**没有贴这种标志的塑料碗**更安全。

Matter can be changed into energy and energy into matter.

(=...and energy can be changed into matter.)

物质可以转化为能,能**也可以转化为**物质。

Under no circumstances can more work be got out of a machine than is put into it.

(=...than work is put into the machine.)

机器输出功决不能大于**输入功**。

This experimental result is in inverse ration to that.

(=...to that experimental result.)

这项实验结果和**那项实验结果**成反比。

When you turn a switch, you can easily operate lighting, heating or power-driven electrical devices.

(=...lighting device, heating device or power-driven electrical device.)

旋动开关,就能轻而易举地启动照明**装置**、供热**装置**和电动**装置**。

Some substances are soluble, while others are not.

(=...while others are not soluble.)

有些物质是可溶的,而另一些是**不可溶的**。

Unit Five　Earth Science

The best conductor has the least resistance and the poorest the greatest.
(=...and the poorest conductor has the greatest resistance.)
最好的导体电阻最小，最差的**导体电阻**最大。

1·5　增加表示时态的词语

汉语的动词没有表示时态的词形变化，因此翻译时应增译相应的时间副词或助词，用来表示不同的时态。如下例。

This natural approach to eliminating hazardous wastes in the soil, water, and air *is capturing* the attention of government regulators, industries, landowners, and researchers.

这种清除土壤、水以及空气中有害废物的自然方法**正日益受到**各政府监管机构、企业、土地所有者和研究人员的关注。

The earth's population *is doubling*, and the environment *is being damaged*.
地球的人口**正在加倍增长**，环境也在不断受到破坏。

It's time to ring farewell to the century of physics and ring in the century of bio-technology.
现在正是告别物理学世纪，迎接生物技术世纪的时候了。

Some day man *will* be able to utilize the solar energy.
总有一天，人类**将**能利用太阳能。

The high-altitude plane *was* and still *is* a remarkable bird.
高空飞机**过去是而且现在仍然还是**一种了不起的飞行器。

Humans *have been dreaming of* copies of themselves for thousands of years.

千百年来，人类**一直**梦想制造出自己的复制品。

This company *has been manufacturing* computers for five years.

这家公司五年来**一直**在生产计算机。

By the turn of the 19th century geologist *had found* that rock layers occurred in a definite order.

到 19 世纪初，地质学家**已经发现**岩层有一定的次序。

1·6 增加表示句子主语的词语

当被动句中的谓语是表示"知道""了解""看见""认为""发现""考虑"等意思的动词时，通常可以在该词前增加"人们""我们""有人"等词语，译为汉语的主动句。如下例。

It is said that numerical control is the operation of machine tools by numbers.

人们说，数控就是用数字对机床加以控制。

It is believed that we shall make full use of the sun's energy some day.

我们相信，总有一天我们将能充分利用太阳能。

It is estimated that the new synergy between computers and Net technology will have significant influence on the industry of the future.

有人预测，计算机和网络技术的新协同作用将会对未来工业产生巨大的影响。

Unit Five　Earth Science

It may be supposed that originally the earth's land surface was composed of rocks only.

我们可以设想地球陆地表面原先完全是由岩石构成的。

Potassium and sodium ***are seldom met*** in their natural state.

我们很少见到自然状态的钾和钠。

The design ***is considered*** practical.

大家认为这一设计切实可行。

1·7　增加具有动作意义的抽象名词

有些英文句子如果直译，意思上就表达得不甚明确。这时，有必要添加一些相关的词语。如下例。

Oxidation will make iron and steel rusty.

氧化**作用**会使钢铁生锈。

The ***lack*** of resistance in very cold metals may become useful in electronic computers, but to keep everything cold, they may have to be placed in liquid helium.

这种在温度极低的金属中没有电阻的**现象**可能对电子计算机很有用处。但若要使每样东西都保持低温，将可能不得不把电子计算机放置在液态氦中。

Were there no electric pressure in a conductor, ***the electron flow*** would not take place in it.

导体内如果没有电压，便不会产生电子流动**现象**。

During *an El Nino* the pressures over Australia, Indonesia and the Philippines are higher than normal, which results in dry conditions or even droughts.

当出现"厄尔尼诺"**现象**时，澳大利亚、印度尼西亚和菲律宾上空的大气压会比平常高，导致出现干旱状况，甚至还会发生旱灾。

The statistics brought out a gender division between hard and soft science: *girls tending toward biology, boys tending maths and physics.*

统计表明，从事硬科学和软科学研究的科学家在性别上存在着差别：女性倾向于生物学的**研究**，而男性则倾向于数学和物理学的**研究**。

1.8 增加表示修辞连贯的连词

由于修辞上的需要，在科技翻译英译汉的过程中增加一些词语可使语气连贯。如下例。

Heat from the sun stirs up the atmosphere, *generating* wind.

太阳发出的热能搅动大气，**于是**产生了风。

In general, all the metals are good conductors, *with* silver the best and copper the second.

一般来说，金属都是良导体，**其中**以银为最好，铜次之。

The pointer of the ampere-hour meter *moves* from zero to two and *goes back* to zero again.

安培小时表的指针**先**从零转到二，**然后**又回到零。

Unit Five　Earth Science

2 减词法

减词法是指在翻译时将原文中的某些词语省略不译的一种方法。省略不译的原因是译文中虽无其词却已有其意，或者是不言而喻的。因此，省略并不是原文的某些思想内容删去，而是一种修辞，是为了达到言简意赅、直截了当的目的。省略的部分一般都是可有可无的，或者是译了反嫌累赘或违背目标语的语言习惯的。

就英译汉而言，由于英语与汉语在句法结构和遣词造句上的差异，英语句子中需要的词，汉语句子中并不一定需要，如果将其译出，反而显得累赘。比如，英语中冠词、介词、连词、代词使用很广泛，但汉语中却根本没有冠词，介词、连词和代词的使用也不是很多。因此在英译汉时，显得累赘、多余的词语往往省略不译，这样译文反而会更加严谨，更加通顺，更加简明扼要。如下例。

This laser beam *covers* a very narrow range of frequencies.

原译：这种激光束的频率覆盖一个很窄的范围。

改译：这种激光束的频率范围很窄。

在例 48 中，"cover"一词在原句中必不可少，但在汉语里将其硬译出来反而累赘。相反，在改译中将"cover"一词省略不译，译文反而显得更加简洁明了。

再如下例。

Heated to a *temperature* of about 500℃ and then quenched, aluminum will develop a great tensile strength.

原译：加热到 500℃左右的温度时再淬火，铝能产生很大的抗拉强度。

改译：加热到 500℃左右再淬火，铝能产生很大的抗拉强度。

在例 49 中，原译将"temperature"一词硬译出来，显得有些啰嗦。在改译中将"temperature"一词省去不译，丝毫没有影响句意的表达。

93

在翻译实践中，要想在忠实于原文的基础上最大限度地追求译文的通顺流畅，常常求助于减词法，科技翻译亦是如此。下面将对减词法的几种形式分别予以介绍。

2·1 省略冠词

英语有冠词，汉语没有冠词。因此，在英译汉时常常将冠词省略。

The chamber formed by *the* top of *the* piston and *the* cylinder head must be virtually air tight for efficient engine operation.

由活塞顶盖和汽缸盖组成的燃烧室必须完全密封，这样发动机才能有效地运转。

在这句话中，有四个名词前面使用了定冠词"the"，但在翻译成汉语时，都可以将它们略去不译。又如下例。

In *a* liquid cooling system, *the* cylinders have an elaborate system of passages called *a* jacket through which cooling fluid is circulated by *a* pump.

在液体冷却系统中，汽缸拥有一系列复杂的通道，称为冷却套；冷却液通过冷却套由泵压产生循环。

在例 51 中，"liquid cooling system" "jacket" "pump" 这三个词语前面的不定冠词"a"和"cylinders"前面的定冠词"the"，在汉译时都应略去不译。

The mechanical energy can be changed back into electrical energy by means of *a* generator or dynamo.

利用发电机能把机械能转变回电能。

该句中可将"mechanical energy"前的定冠词"the"和"generator"前的不定冠词"a"省略不译。

需要指出的是，并不是在所有情况下都可以把冠词省略不译。有的时候，如不定冠词"a"表示数量"一"的时候，定冠词"the"用来特别强调"这个""那个"时，就必须翻译出来。如下例。

The Space Shuttle consists of the fully reusable orbiter vehicle, which has *a* cargo bay designed to carry a variety of payloads.

航天飞机是由一个完全可以重复利用的轨道飞行器组成。轨道飞行器含有一个用来携带各种货物的货舱。

该句中"Space Shuttle"一词前的定冠词"the"可以省去不译，但"cargo bay"前的不定冠词"a"表示数量，必须译出。

The vertical element has *a* fixed part called *the* vertical stabilizer and *a* movable part called *the* rudder.

垂直面上有一个固定的部件，称为垂直安定面，还有一个可移动的部件，称为方向舵。

该句中"vertical element""vertical stabilizer""rudder"等前面的定冠词"the"可以省去不译，但"fixed part"和"movable part"前的不定冠词"a"最好译出来。

2·2 省略代词

可以省略不译的代词有下面几种：

（1）省略句中曾出现过的某一名词的人称代词或指示代词。

The finished products must be sampled to check their quality before *they* leave the factory.

成品在出厂前必须抽样进行质量检查。

The waste gases are harmful to us and we should by all means remove *them*.

废气对我们是有害的，应该尽力加以排除。

Many of the projects or products made in the machine shop have little or no value until *they* are heat-treated.

机器车间生产的许多制品或产品在进行热处理之前几乎没有或完全没有用处。

（2）省略某些作定语的物主代词。

The finished products must be sampled to check *their* quality before they leave the factory.

成品在出厂前必须抽样进行质量检查。

例59

Although alternative iron-making processes are emerging, they are not expected to impact significantly on blast furnace production in the next decade and possibly beyond. Consequently, sintering production should be maintained at *its* present level for some time to come.

虽然出现了可选择的炼铁法，但预计它们在下一个十年或更长的时间内不会对高炉生产产生很大的冲击。因此在将来的一段时期内烧结法生产将保持在目前的水平。

（3）省略主语并含有泛指意思的人称代词"we""you"和不定代词"one"。

Unit Five   Earth Science

You will see to it that the engine doesn't get out of order.

注意别让发动机出毛病。

（4）省略没有实际意义的"it"，如做形式主语或宾语的非人称"it"、强调句中的"it"等。

You will see to *it* that the engine doesn't get out of order.

注意别让发动机出毛病。（形式宾语"it"省略不译）

It was just such a missile, fired by an Iraqi warplane, that nearby sank the USS Stark in the Persian Gulf ten years ago, killing 37 sailors.

十年前，伊拉克战斗机发射的正是这种导弹，在波斯湾附近险些击沉了美国"斯塔克"军舰，炸死了舰上 37 名水手。（强调句中的"it"省略不译）

As the blast furnace is a countercurrent process in which solids descend against a rising gas flow, *it* is imperative that the ferrous burden is supplied in a lumpy form. *It* is necessary, therefore, to agglomerate fine ores by sintering or palletizing.

由于高炉是逆流工作的，在高炉内固体料斗迎着上升的气流而下降，因此炼铁炉料必须以块状形式加入炉内。为此，通过烧结和造球将细粉矿造块是必要的。（"it is imperative that..."和"It is necessary..."中的形式主语"it"省略不译）

Once the desired track is found, the head must be pressed against the disk or loaded. Typically *it* takes about 50 ms to load the head and allow it time to settle against the disk.

一旦找到指定的磁道，磁头必须压在或加载在磁盘上。加载磁头并将其固定在磁盘上一般需要 50 毫秒。

2·3 省略介词

翻译时，有些介词省略不译，仍能完整地表达原文的意义，这样可以做到翻译精练，表意清楚，免去累赘之感。反之，译出每一个介词，汉语句子则显得呆板，甚至不能达到理想的翻译效果。如下例。

Below 4℃, water is *in* continuous expansion instead of continuous contraction.

水在4℃以下就会不断膨胀，而不是不断收缩。

Developers in the US have started work *on* a 100 hp (74.6 kW) motor and hope to have it running *by* the end of next year.

美国的研究人员已经开始研制一种100马力（74.6千瓦）的发动机，并希望明年年底让它运行使用。

Much progress has been made in electrical engineering *in* less than a century.

不到一个世纪，电气工程就取得了很大进展。

On the 1st of January, the company demonstrated a 5 hp (3.7 kW) superconducting motor in action.

1月1日，这家公司现场演示了一台5马力（3.7千瓦）的超导发动机。

2·4 省略连词

英汉两种语言的一个重要区别在于，英语重形合而汉语重意合。在英语文体中，逻辑关系多由连词来表示，相反，在汉语文体中，逻辑关系常由词序和语言内涵来决定，较少借助于连词。因此，在英译汉时，常常省略连词。如下例。

Unit Five Earth Science

As compared to the flooded type systems, there is relatively little liquid refrigerant in the evaporator. The low-pressure gas is drawn through the suction line to the compressor *and* the cycle is repeated.

与满液式系统相比，干式系统蒸发器中的液态制冷剂相对少一些。低压气体由吸气管吸入压缩机中，开始下一次循环。

This kind of machine can work in succession for five *or* six days.

这种机器可以连续运行五六天。

2·5 省略谓语动词

Commercial preservation and transport of food *is* so common that it would be difficult to imagine an unrefrigerated America.

商用食品贮存以及运输在美国极为普遍，人们难以想象若没有制冷这种技术美国会是怎样的。

When the design *is* complete, the system may then be used to produce detailed engineering drawings.

设计完成时，系统就可用于生成详细的工程图。

Quiz

1. Translate the following sentences into Chinese.

    (1) Carbon combines with oxygen to form carbon oxides.

    (2) Contemporary natural science is now working for new important breakthroughs.

    (3) If A is equal to D, A plus B equals D plus B.

    (4) White or shining surfaces reflect heat; dark surfaces absorb it.

    (5) The temperature needed for this processing is lower than that needed to melt the metal.

    (6) Note again that considerable simplification in solving the above can be achieved if the data are made symmetrical.

    (7) Based upon the relationship between magnetism and electricity are motors and generators.

    (8) This experience has been obtained in the realm of far less stringent operational requirements and conditions, particularly in regard to power dissipation.

2. Translate the following sentences into Chinese. While translating, pay more attention to the italicized words.

    (1) *The* motor used to spin *the* floppy disk is usually *a* motor *whose* speed is precisely controlled by negative feedback.

    (2) *The* disks in *a* hard disk system are made of *a* metal alloy, coated on both sides with a magnetic material.

    (3) *It* is also possible to have either *a* centrally located data base directly that is frequently updated, to keep files or records locked, or both, to avoid concurrent updates or interference.

    (4) Our country exploded *its* first atom bomb *in* 1964.

(5) ***The*** nuclear reactions give the sun its constant supply ***of*** energy.

(6) The other financially attractive feature of ***a*** multiminicomputer system is the distribution of real-time operations closer to the user, ***with*** less reliance on ***a*** centralized facility.

(7) Although man's earliest forebears probably knew about, or observed, the effects of cold, ice and snow on their bodies and on things around ***them***—such as the meat they brought home from the hunt—***it*** is not until ***we*** reach the early history of China that we find any reference to the use of these natural refrigeration phenomena to improve the lives of people, and then only for the cooling ***of*** beverages.

(8) Hardening is ***a*** process of heating and cooling steel to increase ***its*** hardness and tensile strength, to reduce ***its*** ductility, and to obtain a fine grain structure. The procedure includes heating the metal above ***its*** critical point or temperature, followed ***by*** rapid cooling.

(9) When steel reaches this temperature—somewhere between 1,400 and 1,600 °F—the change is ideal to make for a hard, strong material if ***it*** is cooled quickly. If the metal cools slowly, it changes back to ***its*** original state.

(10) Carbon and graphite exhibit properties similar to ***those of*** ceramics with two major exceptions. ***They are*** electrically and thermally conductive.

Part B Reading

The Scope of Geology

The world we live in presents an endless variety of fascinating problems which excite our wonder and curiosity. The scientific worker attempts to formulate these problems in accurate terms and to solve them **in the light of** all the relevant facts that can be collected by observation and experiment. Such questions as What? How? Where? and When? challenge him to find the clues that may suggest possible replies. Confronted by the many problems presented by, let us say, an active volcano, we may ask: What are the lavas made of? How does the volcano work and how is the heat generated? Where do the lavas and gases come from? When did the volcano first begin to erupt and when is it likely to erupt again?[1]

Here and in all such **queries** What? refers to the stuff things are made of, and an answer can be given in terms of chemical compounds and elements.[2] The question How? refers to processes—the way things are made or happen or change. The ancients regarded natural processes as **manifestations** of power by irresponsible[3] gods; today we think of them as manifestations of energy acting on or through matter. Volcanic eruptions and earthquakes no longer reflect the **erratic** behavior of the gods of the underworld: they arise from the action of the earth's internal heat on and through the surrounding crust. The source of the energy lies in the material of the inner earth. In many directions, of course, our knowledge is still incomplete: only the first of the questions we have asked about volcanoes, for example, can as yet be satisfactorily answered. The point is not that we now pretend to understand everything, but that we have faith in the orderliness of natural processes. As a result of two or three centuries of scientific investigation we have come to believe that Nature is understandable in the sense that when we ask her questions by way of appropriate observations and experiment, she will answer truly and reward us with discoveries that endure.

Modern geology has for its aim the deciphering of the whole evolution of the earth from the time of the earliest records that can be recognized in the rocks to the present day.[4] So ambitious a program requires much subdivision of effort, and in practice it is convenient to divide the subject into a number of branches. The key words of the three main branches are the materials of the earth's rocky framework (mineralogy and petrology); the geological processes or machinery of the earth, by means of which changes of all kinds are brought about (physical geology); and finally the succession of these changes, or the history of the earth (historical geology).

Geology is by no means without practical importance in relation to the needs and industries of mankind.[5] Thousands of geologists are actively engaged in locating and exploring the mineral resources of the earth. The whole world is being searched for coal and oil and for the ores of useful metals. Geologists are also directly concerned with the vital subject of water supply. Many engineering projects, such as **tunnels, canals, docks** and **reservoirs**, call for geological advice in the selection of sites and materials. In these and in many other ways, geology is applied to the service of mankind.

Although geology has its own laboratory methods for studying minerals, rocks and fossils, it is essentially an open-air science. It attracts its followers[6] to mountains and waterfalls, glaciers and volcanoes, beaches and coral reefs in search for information about the earth and her often puzzling behavior. Wherever rocks are to be seen in **cliffs** and **quarries**, their arrangement and sequence can be observed and their story deciphered. With his hammer and maps the geologist in the field leads a healthy and **exhilarating** life. His powers of observation become sharpened, his love of Nature is deepened, and the thrill of discovery is always at hand.

New Words and Expressions

in the light of 考虑到；按照
query 疑问；问号
manifestation 表明；显露
erratic 古怪的，乖僻的
tunnel 隧道
canal 运河
dock 码头
reservoir 水库
cliff 悬崖，峭壁
quarry 采石场
exhilarating 令人兴奋的

Notes

1. 句中 Confronted by the many problems presented by... 有定冠词 the 的 many problems 既承前，又启后。承前是指前一句提到的 What、How、Where、When；启后是指受到分词短语 presented by an active volcano 的限定，将要提出后面的四个问题。而所提的四个问题又是与前一句一脉相承的。

2. Here and in all such queries What? refers to the stuff things are made of... 句中 things are made of 是一个定语从句，省略了 which 这个在从句中作宾语的关系代词，该从句修饰 the stuff。本句可译为：在所有这些疑问中，"是什么？"这类问题涉及所组成的物质，这可以用化合物和元素来回答。

3. irresponsible 原意是"不负责任的"，这里引申为"为所欲为的""随心所欲的"。

4. Modern geology has for its aim... 本句可理解为：The aim of modern geology is the deciphering of the whole evolution of the earth from the time of the earliest records that can be recognized in the rocks to the present day.

5. by no means without practical importance 是以两个否定来加重肯定的一种表达形式，直译为：绝非毫无重要实际意义。

6. its followers 直译为"地质学的追随者"，这里译为：从事地质学方面工作的人。

Unit Five Earth Science

📝 Exercises

1. Find out the English equivalents of the following Chinese terms from the passage.

   （1）火山   （2）熔岩
   （3）地壳   （4）矿物学
   （5）岩石学   （6）冰河
   （7）珊瑚礁   （8）物理地质学
   （9）地质史学   （10）地球机制

2. Translate the following sentences into Chinese.

   (1) The scientific worker attempts to formulate these problems in accurate terms and to solve them in the light of all the relevant facts that can be collected by observation and experiment.

   (2) The point is not that we now pretend to understand everything, but that we have faith in the orderliness of natural processes. As a result of two or three centuries of scientific investigation we have come to believe that Nature is understandable in the sense that when we ask her questions by way of appropriate observations and experiment, she will answer truly and reward us with discoveries that endure.

   (3) So ambitious a program requires much subdivision of effort, and in practice it is convenient to divide the subject into a number of branches. The key words of the three main branches are the materials of the earth's rocky framework (mineralogy and petrology); the geological processes or machinery of the earth, by means of which changes of all kinds are brought about (physical geology); and finally, the succession of these changes, or the history of the earth (historical geology).

   (4) Geology is by no means without practical importance in relation to the needs and industries of mankind.

   (5) With his hammer and maps the geologist in the field leads a healthy and exhilarating life. His powers of observation become sharpened, his love of Nature is deepened, and the thrill of discovery is always at hand.

Part C Extended Reading

Paleontology

Paleontology is the scientific study of life existent prior to, but sometimes including, the start of the **Holocene Epoch**. It includes the study of fossils to determine organisms' evolution and interactions with each other and their environments (their paleoecology). Paleontological observations have been documented as far back as the 5th century BC. The science became established in the 18th century as a result of Georges Cuvier's work on comparative **anatomy**, and developed rapidly in the 19th century. The term itself originates from Greek: palaios means "old, ancient", on (gen. ontos), "being, creature" and logos, "speech, thought, study".

Paleontology lies on the border between biology and geology, but differs from archaeology in that it excludes the study of **morphologically** modern humans. It now uses techniques drawn from a wide range of sciences, including biochemistry, mathematics and engineering. Use of all these techniques has enabled paleontologists to discover much of the evolutionary history of life, almost all the way back to when Earth became capable of supporting life, about 3,800 million years ago. As knowledge has increased, paleontology has developed specialized **subdivisions**, some of which focus on different types of fossil organisms while others study **ecology** and environmental history, such as ancient climates.

Body fossils and trace fossils are the principal types of evidence about ancient life, and geochemical evidence has helped to **decipher** the evolution of life before there were organisms large enough to leave body fossils. Estimating the dates of these remains is essential but difficult: sometimes **adjacent** rock layers allow radiometric dating, which provides absolute dates that are accurate to within 0.5%, but more often paleontologists have to rely on relative dating by solving the "jigsaw puzzles" of **biostratigraphy**. Classifying ancient organisms is also difficult, as many do not fit well into the Linnaean taxonomy that is commonly used for classifying living organisms and paleontologists more often use cladistics to draw up evolutionary "family trees". The final quarter of the 20th century saw the development of molecular phylogenetics, which investigates how closely organisms are related by measuring how similar the DNA is in their **genomes**. Molecular phylogenetics has also been used to estimate the dates when species **diverged**, but there is **controversy** about the reliability of the molecular clock on which such estimates depend.

New Words and Expressions

paleontology 古生物学
Holocene Epoch 全新世
anatomy 解剖，分解
morphologically 形态学表现为；形态上
subdivision 分支；细分
ecology 生态学

decipher 破译；解读，解释
adjacent 临近的，毗邻的
biostratigraphy 生物地层学
genome 基因组，染色体组
diverge 分歧；偏离
controversy 争议

Exercises

1. **Find out the English equivalents of the following Chinese terms from the passage.**

   （1）古生态学　　　　　　　　（2）比较解剖学
   （3）考古学　　　　　　　　　（4）生物化石
   （5）实体化石　　　　　　　　（6）遗迹化石
   （7）放射性测定年代　　　　　（8）拼图游戏
   （9）林奈分类法　　　　　　　（10）遗传分类学
   （11）家谱　　　　　　　　　　（12）分子系统发生学

2. **Translate the following sentences into Chinese.**

   (1) Paleontological observations have been documented as far back as the 5th century BC. The science became established in the 18th century as a result of Georges Cuvier's work on comparative anatomy, and developed rapidly in the 19th century.

   (2) Use of all these techniques has enabled paleontologists to discover much of the evolutionary history of life, almost all the way back to when Earth became capable of supporting life, about 3,800 million years ago.

   (3) Estimating the dates of these remains is essential but difficult: sometimes adjacent rock layers allow radiometric dating, which provides absolute dates that are accurate to within 0.5%, but more often paleontologists have to rely on relative dating by solving the "jigsaw puzzles" of biostratigraphy.

   (4) The final quarter of the 20th century saw the development of molecular phylogenetics which investigates how closely organisms are related by measuring how similar the DNA is in their genomes. Molecular phylogenetics has also been used to estimate the dates when species diverged, but there is controversy about the reliability of the molecular clock on which such estimates depend.

Unit Six
Life Science (I)

Part A Lecture

被动语态的翻译

汉语和英语都有被动语态，但他们却以不同的形式来表达被动的意义，因此，我们在翻译时要特别注意英汉两种语言被动意义的不同表达法。下面我们主要探讨被动语态的翻译方法，在讲解翻译方法之前让我们先了解被动语态在科技英语中的使用情况。

1 被动语态的使用

被动语态的大量使用是科技英语句法上的一大特征。在科技文章中，大概有三分之一到一半的动词是被动语态。被动语态的使用在下面两个例子中可见一斑。

The Harry Diamond Laboratories performed early advanced development of the Arming Safety Device (ASD) for the Navy's 5-in guided projectile. The early advanced development *was performed* in two phases. In phase 1, the ASD *was designed*, and three prototypes *were fabricated* and *tested* in the laboratory. In phase 2, the design *was refined*, 35 ASD's and a large number of explosive mockups *were fabricated*, and a series of qualification tests *was performed*. The qualification tests ranged from laboratory tests to drop tests and gun firing. The design *was* further *refined* during and following the qualification tests. The feasibility of the design *was demonstrated*.

哈里·戴蒙德实验室对美海军 5 英寸制导炮弹的解除保险装置（ASD）进行了预研。预研工作分为两个阶段进行。第一阶段：先设计出 ASD，并试制三个样件在实验室进行实验；第二阶段，对原设计进行改进并制造出 35 个 ASD 和大量的爆炸模型，接着进行了一系列鉴定试验。试验包括实验室试验、落锤试验和火炮射击试验。在试验期间和试验结束后，又对设计作了进一步的改进。设计方案的可行性已经得到证明。

Unit Six   Life Science (I)

例 2

　　As oil *is found* deep in the ground, its presence *cannot be determined* by a study of the surface. Consequently, a geological survey of the underground rock structure *must be carried out*. If it *is thought* that the rocks in a certain area contain oil, a "drilling rig" *is assembled*. The most obvious part of a drilling rig *is called* "a derrick". It *is used* to lift sections of pipe, which *are lowered* into the hole made by the drilled. As the hole *is being drilled*, a steel pipe *is pushed* down to prevent the sides from falling in. If oil *is struck*, a cover *is* firmly *fixed* to the top of the pipe and the oil *is allowed* to escape through a series of values.

　　石油埋藏于地层深处。因此，仅研究地层表面，无法确定有无石油，必须勘察地下的岩石结构。如果确定了某一区域的岩石层藏着石油，就在此安装钻机。它的主要部分是机架，用以撑架一节一节的钢管，让其下到孔井。一边钻井，一边下钢管，以防周围土层塌陷。一旦出油，就紧固管盖，让油从各个阀门喷出。

　　在例 1 中，12 处谓语动词竟有 9 处采用被动结构。而在例 2 中，14 处谓语动词只有一处为主动结构。在一小段文字中，被动结构的使用尚且如此，在一篇文章中就更可想而知了。

　　为什么在科技英语中经常出现被动语态呢？

　　使用被动语态的第一个原因是：被动语态的意义更清楚。因为科技作者在进行研究和论证时，往往着眼于演绎论证的结果，而将动作的执行者放在次要的位置，这就需要将事物、过程和结果放在句子的中心地位，被动结构正好满足了这一需要。

　　试比较下面两个例子。

　　The gas is carefully heated.
　　气体被小心地加热。

　　He heats the gas carefully.
　　他小心地加热气体。

111

通过对比可知，例3的意义更清楚。

使用被动语态的第二个原因是：句子主语引出最重要的信息。因为主语是句子中非常重要的部分，把重要的信息放在句首当主语，可以引起读者的注意。

试比较下面两个例子。

A long metal bar is fixed in a retort stand by one end. *The other end* is heated in a flame until it becomes red. *The bar and the stand* are then moved away from the heat. *The temperature of the hot end of the bar* is found to fall rapidly.

一根长金属杆的一端被固定在铁架上，另一端在火焰中加热到发红，然后从热源移走金属杆和铁架，发现金属杆受热端的温度迅速下降。

The experimenter fixes a long metal in a retort stand by one end. He heats the other end in a flame until it becomes red. He then moves the bar and the stand away from the heat. He finds that the temperature of the hot end of the bar falls rapidly.

试验者将一根长金属杆的一端固定在铁架上，用火焰加热另一端，直到它发红。然后从热源移走金属杆和铁架。他发现，金属杆受热端的温度迅速下降。

通过对比可知，例5的主语（黑体字）包含了很多信息，更能突出试验的几个步骤。

使用被动语态的第三个原因是：被动句比主动句简短。因为科技文章崇尚准确、严谨和精练，而被动结构在很多情况下正好可以使句子更为紧凑和简短。

试比较下面两个例子。

A barometer is used for measuring atmospheric pressure.

气压计用于测量大气压。

例 8

People use a barometer for measuring atmospheric pressure.

人们用气压计测量大气压。

由此可见，由于使用被动语态可以不提及人，且客观、简明，并能突出最重要信息，所以被动结构也就受到科技工作者的青睐。

2 被动语态的翻译方法

2.1 译成主动语态

上文中我们提到，科技英语最突出的句法特点之一就是大量使用被动语态，而在汉语中，被动句的使用远没有英语广泛，表示被动的手段也十分有限。因此，翻译中往往要对英语的被动句进行反译处理。

英语的被动句一般有以下几种情况：

（1）有些被动句用来表示某种状态或某种动作的结果，动作的实行者不太明确或很难指出。这类句子实际上并不强调被动的意义。

（2）有些被动句意在突出动作的承受者或强调某种行为或动作的结果，而把这些承受者或结果置于主语的位置以示醒目、以便表达。这类句子实际上也并不强调被动的意义。

（3）有些被动句具有典型的被动意义，且带有由 by 引导的行为主体。

（4）有些被动句的表达法是出于行文上的需要。

以上各种被动句均可译成汉语的主动句，而不必、甚至不能把自己局限在原形式里以致译文生硬。具体方法可有以下几种：

▶ 2.1.1 原文主语仍译作主语

不改变主语及句子结构，直接转换成主动语态。这实际上是省略了"被"字的被动句。因为汉语里表示被动意义的句子可以不用"被"字，而直接采用主动语态的形式。

The experiment will be finished in a week.
这项实验将在一周内完成。

The antenna is automatically stabilized in pitch and roll as the airplane changes attitude.
当飞机改变飞行姿态时,天线会随着飞机的俯仰运动自动平衡。

Several elements and compounds may be extracted directly from seawater.
有些元素和化合物可直接从海水中提取。

Pointers are used to build data structures.
指针用来创建数据结构。

While a current is flowing through a wire, the latter is being heated.
电流通过导线时,导线就发热。

▶ 2.1.2 原文其他成分译作主语

将原文主语转译为宾语,原文其他成分改译为主语。

Air can be liquefied by us.
我们能将空气液化。

Only two operations may be carried out on a stack.
存储栈只能进行两种操作。

Unit Six　Life Science (I)

Being very small, an electron cannot be seen by man.

因为电子很小，所以人们看不见。

The non-medical use of certain drugs is forbidden in the United States because they can be dangerous.

美国禁止某些药品用于非医疗方面，因为它们可能会造成危险。

In 1968, the first heart transplant operation was done in South Africa.

1968 年，南非成功进行了第一例心脏移植手术。

▶ 2.1.3 加译一个主语

英语中有些被动句，尤其是带有主语补足语的被动句，要译成汉语主动句，很难保留原句的主语，句中的其他成分也很难改译为主语，这时候常可通过添加"人们""我们""有人""大家"等这样的泛指代词做主语。

When Edison died in 1931, it was proposed that the Americans turn off all power in their homes, streets, and factories for several minutes in honor of this great man.

1931 年爱迪生去世时，有人建议美国人将家里、街道和工厂的电源全部关掉几分钟，以示对这位伟人的悼念。

The causes of air crashes are extensively investigated.

人们对空难事故的原因进行广泛的调查。

During the design and manufacture of an aircraft, great attention is paid to detailed aspects of the design, as well as overall considerations.

在飞机的设计和制造过程中，人们除了要进行总体考虑外，还要高度关注设计上的一些细节。

Proteins, carbohydrates, fats are often grouped together and called organic nutrients.

人们常把蛋白质、糖类和脂肪归为一类，称其为有机养分。

A computer can be given information or orders in various ways.

我们能以各种方式给计算机发出各种信息或指令。

The aerodynamic characteristics of the new missile are examined and compared to a similar conventional missile.

本文研究了这种新型导弹的空气动力特征，并把这些特征与同类常规导弹进行了比较。

In recent years chimpanzees have been taught in laboratory to use sign language for communicating with people.

近年来，有人在实验室里教黑猩猩用手势与人交流。

Steel and its alloys will still be taken as the leading materials in industry for a long time to come.

在今后很长的一段时间里，人们仍将钢以及合金钢当作工业的主要原料。

The United States is often depicted as a nation that has been devouring the world's mineral resources.

人们常把美国说成是一个挥霍世界矿物资源的国家。

2.1.4 译成汉语的无主句

如果原被动句中的主语或其他成分不宜译作汉语主动句中的主语，而加译一个主语又无必要，那么可将句子反译成汉语的无主句。译时可在原主语前加译"把""将""给""对"等词，从而把原主语译成宾语；也可不加任何介词直接译成宾语。

If water is heated, the molecules move more quickly.

如果把水加热，分子运动就更快。

Attempts were made to regenerate the catalyst by passing oxygen at the rate of 8 cc/sec for 4 hours at 300 ℃.

曾试图再生催化剂，其方法是在 300℃以下以 8 立方厘米每秒的流速通氧 4 小时。

Because of this, applications such as high-temperature wires, heat sinks, and continuous casting mold are foreseen.

因此，可预见到像高温电线、散热器和连铸模具一类的用途。

Air resistance must be given careful consideration when the missile is to be designed.

设计导弹时，必须仔细考虑空气阻力。

The degree of water pollution can be detected with the apparatus newly invented.

采用这种新发明的仪器可以检测出水质污染的程度。

When a leaf is looked at under a microscope, it is seen to have thousands of little breathing pores.

如果把一片树叶置于显微镜下观察，就会看到数千个微小的气孔。

Different stars may be seen rising into view at different times of the year.

在一年的不同时间，可以看到在天空中升起的不同的星体。

2.1.5 以 it 做形式主语的被动句的反译处理

在科技英语文章中，以"it + 被动态谓语 + 主语从句"的句子非常多。翻译时，it 应一律减去不译，而把被动态谓语动词进行拆译处理，整个句子常译成汉语的无主句或有主语的主动句。

It was thought at one time that compounds of carbon were only produced in living organisms.

人们一度认为只有生物体才会产生碳的化合物。

It has been found that some variations can be passed on from one generation to another and that others cannot.

已经发现，有些变异可以遗传，有些不可以。

It is estimated that the human eye can distinguish 10 million different shades of colors.

据估计，人类的眼睛可区分 1,000 万种不同的色调。

It is generally believed that oil originates in marine plant and animal life.

人们普遍认为，石油来源于水生动植物。

Unit Six　Life Science (I)

It has been predicted that there will be an earthquake here in a few days.

据悉，几天内这里将发生地震。

It is claimed that this natural organic compound can be synthesized by artificial process.

有人宣称，这种天然有机化合物可以通过人工合成。

2.2　保持被动结构

英语的被动由 "be + 过去分词" 构成，而汉语则是依靠一些表示被动含义的词来表示。因此，我们需要借助一些词来进行英汉间的转换。

2.2.1 使用 "被" "由" "受" "用" 等字

With afterburners, fuel is injected behind the combustion section and ignited to increase thrust at the expense of high fuel consumption.

通过再燃烧装置，燃料被注入燃烧区后面，然后点燃，由大量燃烧来增加动力。

A modem is used to communicate with another computer over telephone line.

通过电话线，"猫"（调制解调器）被用来连接不同的电脑。

People with allergic disease are particularly hard hit by cigarette.

患过敏疾病的人，特别容易受到香烟的侵害。

Everything on or near the surface of the earth is attracted by the earth.
地球上或地球附近的一切物体都受地球吸引。

It is of logic circuit that all computers are made.
所有的计算机都由逻辑电路组成。

2.2.2 使用"加以""予以""为……所"等字

Other advantages of our invention will be discussed in the following.
本发明的其他优点将在下文中予以讨论。

The finished products must be carefully inspected before delivery.
成品在出厂前必须仔细地予以检查。

This extraction rate was confined in batch tank tests.
这一提取速率已为分批箱内试验所证实。

These problems must be solved before the tests start.
这些问题必须在试验开始前加以解决。

2.2.3 译成"是……的"句型

Iron is extracted from the ore by smelting in the blast furnace.
铁是通过高温冶炼从矿石中提取的。

Unit Six Life Science (I)

Usually the dues are calculated on the registered tonnage of the ship.
通常港口税是以船只的注册吨位计算的。

Malaria is caused by a tiny parasite carried by mosquitoes.
疟疾是蚊子携带的极小的寄生虫传染的。

Some people have assumed that planets and meteors were formed in the same astronomic event.
有人设想行星和流星都是在同一次天体演变中形成的。

▶ 2.2.4 译为"被……为""被……是"形式的正规被动句

在科技英语中，被动结构后由 as 引导出主语补足语的情况很多，我们可将其固定译为"被……为""被……是"形式的正规被动句。

The heart is usually considered as the most important organ of the body.
心脏往往被看作是人体最重要的器官。

由名词（或名词词组）做主语补足语的被动句也可以这样译出。

Radio and television are generally considered the two most efficient mass media.
广播和电视通常被视为效率最高的两种大众传媒。

Poor conductors of heat are often called heat insulators.
不良导热体常被称为热的绝缘体。

2.3 谓语分译

上述各种译法都将被动语态的谓语动词包含在原来的句子内，而谓语分译这种方法则是将被动语态从原句中分出来，译成带主语（泛指人称的）或不带主语的独立成分。

Hydrogen is known to be the lightest element.

人们知道，氢是最轻的元素。

It is just the energy which the atom thus yields up that is held to account for the radiation.

人们认为，这种辐射正是由原子释放出来的能量造成的。

Quiz

1. Translate the following paragraph into Chinese. While translating, pay more attention to the italicized words.

   Almost everyone *is involved* with design in one way or another, even in daily living, because problems *are posed* and situations arise which *must be solved*. A design problem is not a hypothetical problem at all. Design has an authentic purpose—the creation of an end result by taking definite action, or the creation of something having physical reality. In engineering, the word design conveys different meanings to different persons. Some think of a designer as one who employs the drawing board to draft the details of a gear, clutch, or other machine member. Others think of design as the creation of a complex system, such as a communications network. In some areas of engineering the word design *has been replaced* by

Unit Six Life Science (I)

other terms such as systems engineering or applied decision theory. But no matter what words *are used* to describe the design function, in engineering it is still the process in which scientific principles and the tools of engineering—mathematics, computers, graphics, and English—*are used* to produce a plan which, when carried out, will satisfy a human need.

Part B Reading

History of Biology

Contributions to the development of biology have come from all over the world. Three groups of biologists, working in the years since the Renaissance, will be studied here.[1] Their contributions are important to the history of biology, and to modern science. The work of one man in each of the three groups will be studied.

The first group of biologists, and the earliest group to be studied here, are the microscopist of the 17th century. These people worked with microscopes. They built them and improved them for use in the study of science. Anthony van Leeuwenhoek,[2] a Dutch microscopist who lived from 1632 until 1723, was one of the many important people in this group. Van Leeuwenhoek was interested in improving the lenses that were used in making microscopes. He made some microscopes, and looked at many different things with the help of the magnifying lenses which he made, also. By looking through the lenses, van Leeuwenhoek realized that there was a whole world filled with **microscopic** living things. Most people were unaware that these small living things existed. The microscope continues to be a very important tool in science today.

A second group of biologists worked in the 18th century to systematize our knowledge in science.[3] They tried to organize all of the information found by many scientists so that everyone could use the same system for talking about discoveries.[4] One system was developed by a Swedish scientist named Carl von Linnaeus[5], who lived from 1707 until 1778. He classified plants, animals and minerals in a very useful way. His idea was to give each plant, animal, and mineral a two-part Latin name. The first part of the name was a general name. It told what general group of things the plant, animal, or mineral belonged to. This was the name of **genus** or

Unit Six   Life Science (I)

group.[6] The second part of the name was the specific name. This was the name of the species, or kind.[7] It told what specific plant, animal, or mineral it was.

This system was extremely popular among scientists, and is still used today. There are several reasons for its popularity. First, the system is simple and clear. Second, Linnaeus used Latin words in his system and, at that time, nearly all scientists knew Latin. Everyone who knew Latin did not have to learn any special words. Also, the two names were a short description and were fairly easy to remember. Linnaeus's system, which is still used today, is sometimes referred to as a system of **binomial nomenclature**.

A third group of biologists did most of their work in the 19th century. These scientists profited from the interest in world exploration during this time. They went on many expeditions as observers and collectors. Their job was to study the plants and animals of the new lands. One of the best known explorers and observers was the great English biologist Charles Darwin[8]. He lived in England in the years from 1809 until 1882.

Since he was an explorer, Darwin did not spend all of the years of his life at home in England.[9] He left England for five years in the early 1830s to travel on a ship called **the Beagle**. This trip is famous. For the other people on the Beagle, the purpose of this trip was to draw maps and to explore South America. They also planned to sail all the way around the earth. For Darwin, the purpose of the trip was different. He collected many samples of plants and animals from South America and nearby Seas. He also wrote down many of his observations of the living things he found in his explorations. When he returned to England, Darwin wrote a book called *Origin of Species*[10], which was about **evolution**. His theory of evolution was developed as a result of his observations during his trip on the Beagle.

Van Leeuwenhoek, Linnaeus, and Darwin are three very important men in the history of biology. Each is one of a group of people who made a significant contribution to science. These three men have made important contributions to science, but they are only a few of the important people in the history of biology.[11]

New Words and Expressions

microscopic 微观的，用显微镜可见的
genus（种）类，属
binomial 双名的
nomenclature（分类学上的）命名法
the Beagle 小猎兔犬号，密探号
evolution 进化，演变

Notes

1. 单词 group 意为"群；团体；类"等，句中 group 意为"类"；three groups of biologists 指"三种不同类型的生物学家"。句中 will be studied here 译为：这里将研究……，本文将研究……。

2. Anthony van Leeuwenhoek 安东尼·范·拉乌文胡克（1632—1723），荷兰显微镜学家。

3. 句中 a second 表示"再一、又一"的意思，a second group of biologists 意为"再一类/又一类生物学家"。下文中出现的 a third group of 表示"还有一类"的意思。

4. 句中 system 意为"分类法"。全句译为：他们试图把由众多科学家发现的全部知识都加以系统化，以便使大家在谈论科学发现时，都能使用相同的分类法。

5. Carl von Linnaeus 卡儿·冯·林奈乌斯（1707—1778），瑞典科学家。

6. 注意前面两句中的术语：general name 通称；general group 大类；the name of genus or group 类属名称。

7. 注意前面两句中的术语：specific name 种名；the name of the species, or kind 物种名称或品种名称。

Unit Six　Life Science (I)

8. Charles Darwin 查尔斯·达尔文，英国生物学家。
9. 句中 not...all... 构成部分否定。全句译为：作为一名探险家，达尔文没有在英格兰家里度过他的一生。
10. *Origin of Species* 意为《物种起源》。
11. 句中 only 修饰动词，全句译为：他们三位对科学都做出了重大的贡献，但他们仅仅是生物学发展史上重要人物中的几位代表而已。

Exercises

1. **Find out the English equivalents of the following Chinese terms from the passage.**

   （1）生物学家　　　　　　　（2）显微镜学家
   （3）文艺复兴（时期）　　　（4）显微镜
   （5）放大镜（凸透镜）　　　（6）通称
   （7）类属名称　　　　　　　（8）种名
   （9）物种名称或品种名称　　（10）对科学做出贡献

2. **Translate the following sentences into Chinese.**

   (1) Contributions to the development of biology have come from all over the world. Three groups of biologists, working in the years since the Renaissance, will be studied here.

   (2) By looking through the lenses, van Leeuwenhoek realized that there was a whole world filled with microscopic living things.

   (3) Linnaeus's system, which is still used today, is sometimes referred to as a system of binomial nomenclature.

   (4) When he returned to England, Darwin wrote a book called *Origin of Species*, which was about evolution. His theory of evolution was developed as a result of his observations during his trip on the Beagle.

   (5) Van Leeuwenhoek, Linnaeus, and Darwin are three very important

men in the history of biology. Each is one of a group of people who made a significant contribution to science. These three men have made important contributions to science, but they are only a few of the important people in the history of biology.

Part C Extended Reading

Branches of Biology

Brief introduction to biology

Biology is the science that deals with living things. It is broadly divided into botany, the study of plant life, and zoology, the study of animal life. Subdivisions of each of these sciences include cytology (the study of cells), histology (the study of tissues), anatomy or morphology, physiology, and embryology (the study of the embryonic development of an individual animal or plant). Also included in biological studies are the sciences of genetics, evolution, paleontology, and taxonomy or systematics, the study of classification. The methods and attitudes of other sciences are brought to the study of biology in such fields as biochemistry (physiological chemistry), biophysics (the physics of life processes), bioclimatology and biogeography (ecology), bioengineering (the design of artificial organs), **biometry** or biostatistics, bioenergetics, and biomathematics. Evidences of early human observations of nature are seen in prehistoric cave art. Biological concepts began to develop among the early Greeks. The biological works of Aristotle include his observations and classification of his large collections of animals. The invention of the microscope in the 16th century gave a great stimulus to biology, broadening and deepening its scope and creating the sciences of microbiology, the study of microscopic forms of life, and microscopy, the microscopic study of living cells. Among the many who contributed to the science are Claude Bernard[1], Cuvier[2], Darwin, T. H. Huxley[3], Lamarck[4], Linnaeus, Mendel[5], and Pasteur[6].

Botany

Botany, the study of plants, occupies a peculiar position in the history of human knowledge. For many thousands of years it was the one field of awareness about which humans had anything more than the vaguest of insights. It is impossible to know today just what our Stone Age ancestors knew about plants, but from what we can observe of pre-industrial societies that still exist, a detailed learning of plants and their properties must be extremely ancient. This is logical. Plants are the basis of the food pyramid for all living things, even for other plants. They have always been enormously important to the welfare of peoples, not only for food, but also for clothing, weapons, tools, dyes, medicines, shelter, and a great many other purposes. Tribes living today in the jungles of **the Amazon** recognize literally hundreds of plants and know many properties of each. To them botany, as such, has no name and is probably not even recognized as a special branch of "knowledge" at all.

Unfortunately, the more industrialized we become the farther away we move from direct contact with plants, and the less distinct our knowledge of botany grows. Yet everyone comes unconsciously on an amazing amount of botanical knowledge, and few people will fail to recognize a rose, an apple, or an **orchid**. When our Neolithic ancestors, living in the Middle East about 10,000 years ago, discovered that certain grasses could be harvested and their seeds planted for richer yields the next season, the first great step in a new association of plants and humans was taken. Grains were discovered and from them flowed the marvel of agriculture: cultivated crops. From then on, humans increasingly take their living from the controlled production of a few plants, rather than getting a little here and a little there from many varieties that grew wild—and the accumulated knowledge of tens of thousands of years of experience and intimacy with plants in the wild would begin to **fade away**.

Zoology

Zoology is concerned with the study of animal life. From earliest times animals have been vitally important to man; cave art demonstrates the practical and mystical significance animals held for prehistoric man. Early efforts to classify animals were based on **physical resemblance**, **habitat**, or economic use. Although Hippocrates and Aristotle did much toward organizing the scientific thought of their times, **systematic** investigation declined under **the Romans** and, after Galen's notable contributions, came to a virtual halt lasting through the Middle Ages (except among the Arab physicians). With the Renaissance direct observation of nature revived; landmarks were Vesalius'[7] anatomy and Harvey's[8] demonstration of the **circulation of blood**. The invention of the microscope and the use of experimental techniques expanded zoology as a field and established many of its branches, e.g., cytology and histology. Studies in embryology and morphology revealed much about the nature of growth and the biological relationships of animals. The system of binomial nomenclature was devised to indicate these relationships; Linnaeus was the first to make it consistent and apply it systematically. Paleontology, the study of fossil organisms, was founded as a science by Cuvier in 1812. Knowledge of physiological processes expanded greatly when physiology was integrated with the chemical and other physical sciences. The establishment of the cell theory in 1839 and the acceptance of **protoplasm** as the stuff of life 30 years later gave impetus to the development of genetics. Lamarck, Mendel, and Darwin presented concepts that revolutionized scientific thought. Their theories of evolution and of the physical basis of heredity prompted research into all life processes and into the relationships of all organisms. The classic work of Pasteur and Koch[9] opened up bacteriology as a field. Modern zoology has not only concentrated on the cell, its parts and functions, and on expanding the knowledge of cytology, physiology, and biochemistry, but it has also explored such areas as psychology, anthropology, and ecology.

New Words and Expressions

biometry 寿命测定；生物统计学
the Amazon 亚马孙河
orchid 兰花
fade away 渐渐消逝
physical resemblance 形态相似
habitat 栖息地
systematic 系统化的
the Romans 罗马人
circulation of blood 血液循环
protoplasm 原生质；细胞质

Notes

1. Claude Bernard 克劳德·伯纳德，法国生物学家。
2. Cuvier 居维叶，法国古生物学家。
3. T. H. Huxley 赫胥黎，英国生物学家。
4. Lamarck 拉马克，法国博物学家。
5. Mendel 孟德尔，奥地利生物学家和遗传学家。
6. Pasteur 巴斯德，法国化学家和细菌学家。
7. Vesalius 维萨里，比利时解剖学家。
8. Harvey 哈维，英国生理学家和医生。
9. Koch 科赫，德国细菌学家和医生。

Exercises

1. Find out the English equivalents of the following Chinese terms from the passage.

   （1）细胞学　　　　　　　　（2）组织学
   （3）解剖学　　　　　　　　（4）形态学
   （5）生理学　　　　　　　　（6）胚胎学
   （7）遗传学　　　　　　　　（8）古生物学
   （9）分类学　　　　　　　　（10）生理化学
   （11）生物气候学　　　　　（12）生物地理学
   （13）生态学　　　　　　　（14）生物工程学
   （15）生物统计学　　　　　（16）生物能量学
   （17）微生物学　　　　　　（18）心理学
   （19）细菌学　　　　　　　（20）人类学

2. Translate the following sentences into Chinese.

   (1) Biology is the science that deals with living things. It is broadly divided into botany, the study of plant life, and zoology, the study of animal life.

Unit Six Life Science (I)

(2) Evidences of early human observations of nature are seen in prehistoric cave art.

(3) The invention of the microscope in the 16th century gave a great stimulus to biology, broadening and deepening its scope and creating the sciences of microbiology, the study of microscopic forms of life, and microscopy, the microscopic study of living cells.

(4) Botany, the study of plants, occupies a peculiar position in the history of human knowledge. For many thousands of years it was the one field of awareness about which humans had anything more than the vaguest of insights.

(5) To them botany, as such, has no name and is probably not even recognized as a special branch of "knowledge" at all.

(6) Unfortunately, the more industrialized we become the farther away we move from direct contact with plants, and the less distinct our knowledge of botany grows.

(7) When our Neolithic ancestors, living in the Middle East about 10,000 years ago, discovered that certain grasses could be harvested and their seeds planted for richer yields the next season, the first great step in a new association of plants and humans was taken.

(8) From then on, humans increasingly take their living from the controlled production of a few plants, rather than getting a little here and a little there from many varieties that grew wild—and the accumulated knowledge of tens of thousands of years of experience and intimacy with plants in the wild would begin to fade away.

(9) Knowledge of physiological processes expanded greatly when physiology was integrated with the chemical and other physical sciences.

(10) Modern zoology has not only concentrated on the cell, its parts and functions, and on expanding the knowledge of cytology, physiology, and biochemistry, but it has also explored such areas as psychology, anthropology, and ecology.

Unit Seven
Life Science (II)

Part A　Lecture

定语从句的翻译

英语的定语从句是各种从句中最复杂、最难译的一种。它是由关系代词或关系副词引导的从句，在句中作定语，一般是对某一名词或代词进行修饰和限制，被修饰的词称为先行词，而定语从句位于先行词之后。

1　英汉定语结构的对比

一般来说，英语的定语从句相当于汉语中作定语的动宾结构、动补结构、偏正结构、主谓结构等。但实际上，英语的定语从句要比汉语的这几种定语结构复杂。一方面，汉语的定语总是位于其修饰的词之前，少用或忌用长的定语而保持句子意义上的平衡。而英语的定语从句总是位于所修饰的词之后，且都是完整的主谓结构，以保持其形式上的平衡。另一方面，汉语的定语往往只起修饰限制的作用，而英语的定语从句则不然，它虽然在形式上只是起定语的作用，但是在实际操作中，它却远远超出了定语的范围，可以进行补充说明、交代细节、分层叙述以及表示条件、原因、目的、结果或转折等意义。可以说，英语中的定语从句的范围远远大于汉语定语的范围，从这个意义上讲，要将定语从句翻译好恐怕不是一件轻而易举的事。

另外，英语的定语从句在结构形式上有许多种，如直接由关系代词、关系副词引导的，由"介词＋关系代词"引导的，由"名词＋介词＋关系代词"引导的，由"介词＋which＋名词"引导的，以及许多割裂式的定语从句。英语定语从句结构上的复杂，在一定程度上增加了理解和翻译的难度。

2　定语从句的翻译方法

英语定语从句的译法主要涉及限制性定语从句和非限制性定语从句的译法。从理论上讲，这两类定语从句有着本质的区别，因此翻译方法也应有所不同。但在实际语言应用过程中，这两类从句交叉出现的现象非常普遍，因此其翻译处理方法并没有明显的区别。实践表明，定语从句的译法主要取决于句子的长短和复杂程度。

2·1 限制性定语从句的译法

大部分限制性定语从句起着限制先行词的作用，无论从结构还是意义上讲都是全句中必不可少的。因此翻译时可将从句融合在主句中，译作先行词的前置定语。这样译时，一般要把关系代（副）词减去不译。试看以下各例。

All the plants and animals which we know of have to breathe and so can live only on planets which have suitable atmosphere.

我们所知的一切动植物都必须呼吸，因而只能生存在有适宜大气层的星球上。

Language is a tool by means of which people communicate ideas with each other.

语言是人们赖以交流思想的工具。

Surface subsidence was most serious in certain areas where large quantities of underground water were pumped out.

在大量抽取地下水的某些地区，地面下沉非常严重。

PCTOOLS are tools whose functions are very advanced.

PCTOOLS 是功能很先进的工具。

The man who is making the experiment graduated from the University of California, Berkeley.

正在做实验的这个人是加州大学伯克利分校毕业的。

从理论上讲，限制性定语从句应该是全句不可缺少的成分，但实际上并不总是如此。有不少限制性定语从句和主句之间虽没有逗号隔开，但实质上对先行词并没有明显的限制性，单独译出并不会破坏主句的完整性。还有一些限制性定语从句较长，如采用逆序可能造成修饰语太长或层次不清。在这两种情况下，都可以采用顺序分译成并列句的方法处理。限制性定语从句采用顺序分译法非常普遍，甚至不亚于非限制性定语从句，这在很大程度上是由于汉语多用短语的习惯造成的。这样翻译时对关系代（副）词的处理方法有三种：一是减译关系代（副）词而重复先行词；二是用代表先行词的代词取代关系代（副）词；三是减去关系代（副）词不译。

The signals are reports from the senses, mostly the ears and eyes, which keep the brain in touch with the world around.

这些信号是由感觉器官（主要是耳朵和眼睛）传来的报告，它们保持大脑与周围世界的联系。（用代词代替关系代词）

The most ambitious project so far is a transatlantic fiber-optic cable to be built by 1988 that could significantly cut the cost of communication between the United States and Europe.

迄今为止，最为雄心勃勃的工程是将于1988年铺成的横贯大西洋的光纤电缆线路，这条线路将大大减少美国和欧洲之间的通信成本。（重复先行词）

The electricity is changed into the radio-frequency power which is then sent out in form of radio waves.

电转变成射频能，接着以无线电波的形式发射出去。（减译关系代词，译为承前省略句）

2·2 非限制性定语从句的译法

有些定语从句从形式上看是非限制性的，但从内容上看却是限制性的。

Unit Seven　Life Science (II)

因此，这类非限制性定语从句有时也可以采用逆序译成定语的方法译出，以避免译文句子结构松散，语气不连贯。

Our cities have many factories, which we need to make food products, clothing, many other things.

城市里有许多加工食物、生产衣服和其他产品的工厂。

For pyrite and pyrrhotite concentrates, which have had limited commercial value to date, the desulphurization process offers an economic means of producing both elemental sulphur and high-grade iron oxide.

对于迄今为止仅具有有限工业价值的黄铁矿精矿和磁黄铁矿精矿而言，这种脱硫法提供了同时生产元素硫和高等级氧化铁的经济手段。

A prism, which is made of glass, can break up a beam of incoherent light.

玻璃棱镜能分解非相干光。

Perhaps this is the "dead ray", which we often read about in science fiction.

这也许就是我们常在科幻小说中读到过的"死亡射线"。

The electrons, with their negative charges, revolve about the nucleus, which is always positively charged.

带负电的电子，绕着带正电子的核旋转。

但是大部分非限制性定语从句都是对先行词的解释或补充说明，两者之间的关系并不十分密切，从句在意义上具有独立性，实际上相当于和主句并列的分句，对整个句子来说并非必不可少。因此，这种定语从句往往可以和主语分开来译，译在其先行词之后，作为和主句并列的一个分句。具体方法和对限制性定语从句的处理相同。

例14

All matter is made of atoms, which are too small to be seen even through the most powerful microscope.

一切物质皆由原子构成。原子本身太小，即使用最大倍数的显微镜也无法看到。（重复先行词）

例15

This was the beginning of the science of radar, which finds aircraft by the reflections of radio waves sent into the sky.

这就是雷达科学的开端。雷达利用射入天空中的无线电波的反射波来发现飞机。（重复先行词）

例16

The properties of carbon steels, which are most widely used, depend on the amount of carbon contained.

碳钢的用途十分广泛，其性能取决于含碳量的多少。（用代词代替关系代词）

例17

When oil wells are drilled, the first material obtained is frequently natural gas, which burns with a hot flame.

钻井时，首先获得的物质常常是天然气，它燃烧时发出炙热的火焰。（用代词代替关系代词）

例18

The steam travels along pipes to a turbine, where it drives the shaft at high speed.

蒸汽由管道传送到汽轮机，再以高速度驱动汽轮机主轴。（减译关系副词，从句译为承前省略句）

例19

The last big Alaska earthquake created a tsunami, that was felt 1,500 miles away.

最近发生的阿拉斯加大地震引起了海啸，在 1,500 英里之外都能感觉到。（减译关系代词，从句译为承前省略句）

2.3 兼有状语功能的定语从句的译法

英语的定语从句和主句之间的关系很复杂，有时仅从语法结构上分析从句和先行词的关系会给理解和翻译造成困难。这是因为有一些定语从句（包括限制性和非限制性定语从句）对先行词的修饰限制作用已经很弱，而相当于状语的作用，往往含有原因、条件、结果、让步、目的、时间等意义。因此，翻译时应仔细分析主句和从句之间的逻辑关系，把具有状语职能的定语从句转换成适当的状语译出。

▶ 2.3.1 译为原因状语

In the place beyond an object that the light cannot directly reach a shadow is formed.

在物体背后因为光线不能直接到达而形成影子。

The instruments that are light in weight and small in volume are used in this space shuttle.

这批仪器重量轻，体积小，故用于这架航天飞机上。

All planets but the earth that are very far from the sun and extremely cold are not places where livings can ever exist.

除了地球以外，在所有的行星生物都不能生存，因为它们离太阳非常远而且极为寒冷。

Copper whose resistance is low can serve as a good conductor.

铜因为电阻小，所以是一种良导体。

2.3.2 译为结果状语

The moon is an airless and waterless waste where there is no life of any sort.

月球是一个无空气、无水的荒地,所以那里没有任何的生物。

All matter has certain features or properties that enable us to recognize it from chemical and physical tests.

所有物质都有某些特征或特性,这使得我们可用化学和物理试验来鉴别它们。

It is fortunate that men have worked out new plane shapes which enable the plane to go through sound barrier with little difficulty.

幸而,人们已经设计出新的机型,使飞机通过音障已不太困难。

2.3.3 译为目的状语

People used to have earthquake explorations which found petroleum structures under the sea bed.

过去人们都用地震勘探来寻找海底石油构造。

An improved design of radiators must be made which results in more uniform temperature distributions, at least, if not raise temperature.

必须改进散热器的设计,使之即使不能提高温度,也能使温度分布地更均匀。

Unit Seven　Life Science (II)

Petroleum must be moved to a refinery where these compounds can be separated.

必须把石油输送到炼油厂，以便把这些化合物分解。

▶ 2.3.4 译为让步状语

Potential energy that is not so obvious as kinetic energy exists in many things.

势能虽不如动能那么明显，但它存在于许多物体之中。

Electrons, which normally repel each other, seem to travel in pair at unusually cold temperature.

电子虽然在常态下相互排斥，但在极低的温度下似乎是成对地运动。

Photographs are taken of stars and other objects, the light of which is too faint to be seen by eyes at all.

虽然许多星球和其他天体的光线非常微弱，眼睛根本看不见，但它们的照片还是被拍下来了。

▶ 2.3.5 译为条件状语

Substances that contain only atoms with same properties are called elements.

如果物质只包含相同性质的原子，就称为元素。

143

A body whose position changes with time is said to be moving.

如果物体的位置随时间而变化，就认为它在运动。

In hospitals where computers are used, diagnosis becomes quicker and more accurate.

如果医院里用了计算机，诊断就会更快，更准确。

2.3.6 译为时间状语

The quantity of fluid which passes through a given section of the pipe can be measured.

当流经已知管道截面时，可以测得其流量。

The object whose weight is more than that of the water displaced will sink.

当物体的重量大于它排开的水的重量时，它就会下沉。

Electrical energy that is supplied to a lamp can be turned into light energy.

给电灯供电时，电能就变成光能。

2.4 割裂式定语从句的译法

英语中的定语从句大都紧跟在其先行词之后，但有时也被其他成分隔开，形成割裂的现象。产生这种现象的原因，大多是句子结构本身要求造成的。在下述情形中，定语从句被迫离开先行词而造成割裂修饰现象：(1) 先行词带介词短语、分词短语、形容词短语、同位语短语和非限制性定语从句作后

Unit Seven　Life Science (II)

置定语时,定语从句被迫离开先行词;(2)当主句的谓语很短时,为使谓语紧接在主语之后,修饰主语的定语从句被迫离开先行词。对这种定语从句的翻译,关键是要从语法分析和逻辑分析入手,正确地判断出其先行词,然后再根据从句与先行词之间的关系和整个句子的意思,灵活地采用上述几种译法加以处理。

There is a space beyond the object that the light cannot reach directly.

在物体的另一面有一个光线不能直接到达的地方。(定语从句修饰 a space)

The forces acting on a given body which are exerted by other bodies are referred to as external forces.

由别的物体作用于已知物体上的力称为外力。(定语从句修饰 the forces)

例41

Although television was developed for broadcasting, many important uses have been found that have nothing to do with it.

虽然电视是为广播而研制的,但它还有许多与广播无关的重要用途。(定语从句修饰 many important uses)

例42

An order has come from Berlin that no language but German may be taught in the school of Alsace and Lorraine.

从柏林下达了一道命令,在阿尔萨斯和洛林地区的学校除了德语以外,不准教其他语言。(定语从句修饰 an order)

Whenever one surface moves over another, a force is set up which resists the movement.

每当一个平面沿着另一个平面运动时,就会产生一个阻碍这种运动的力。(定语从句修饰 a force)

Is the earth the only heavenly body in the whole enormous universe where human beings or anything like human beings exist?

在广袤的宇宙中,难道只有地球这个天体才有人类或类似于人类的生物存在吗?(定语从句修饰 the only heavenly body)

This led to a point of view, which has often been suggested, that the sea-keeping quality of a vessel may be determined from a single sea state.

人们常提出这样一个观点,即船舶适航性可由一种海况来决定。(定语从句修饰 a point of view)

2.5 特殊定语从句的译法

英语的定语从句,有三种比较特殊,一种是由 which 或由介词加 which 引导的,用来修饰主句部分或全部内容的非限制性定语从句;一种是由 as 引导的,也是用于说明整个主句内容的非限制性定语从句;还有一种是由"名词(代词、数词)+ of + which"引导的定语从句。它们的翻译方法将专门进行讨论。

▶ 2.5.1 由 which 引导的定语从句

由 which 引导的非限制性定语从句可用来修饰整个主句或主句的部分内容。它具有状语的功能,起补充说明、强调、对比的作用,还可表示原因、结果、目的、让步及条件等。翻译时,把它当作一个并列分句或状语从句来处理,并适当加上"因为""如果""由于""虽然"等连词。

Metals can conduct electricity, which plastics cannot.

金属能导电,而塑料却不能。(对比)

Unit Seven Life Science (II)

Like charges repel and opposite charges attract each other, which is one of the fundamental laws of electricity.

同类电荷相斥,异类电荷相吸,这是电学中的基本定律。(强调)

One day on the moon is as long as two weeks on the earth, which has been proved by astronomers.

月球上的一天有地球上的两个星期那么长,这已为天文学家所证实。(补充和说明)

All forces occur in pairs, which may be spoken of as the principle of action and reaction.

所有的力都成对出现,这就是作用与反作用原理。(补充和说明)

Moist atmosphere makes iron rust rapidly, which leads us to think that water is the influence causing the corrosion.

潮湿的大气使铁很快生锈,这会使我们认为水是引起腐蚀的原因。(表示结果)

Copper, which is used widely for carrying electricity, offers very little resistance.

铜的电阻很小,故广泛用于传送电力。(表示结果)

The mass of a body differs from its weight, which varies with position and elevation.

物体的质量和重量是不同的,因为重量随其所在的位置和高度而变化。(表示原因)

The center of gravity of a passenger ship should be as low as possible, the initial stability of which may be greatly increased.

客船的重心应尽可能低,以便极大提高其初始稳定性。(表示目的)

Diagnoses are made quick and accurate by an electronic computer, the use of which is introduced in the hospital.

如果医院采用电子计算机,诊断会变得更快,更准确。(表示条件)

Friction, which is often considered as a trouble, is sometimes a help in the operation of machine.

虽然人们常认为摩擦是一种麻烦,但它有时却有助于机器的运转。(表示让步)

2.5.2 由 as 引导的定语从句

由 as 引导的非限制性定语从句通常修饰整个主句或主句的部分内容,引导词 as 必须充当从句的主语。这种从句可置于句子的首位,或夹在中间,前后用逗号隔开,构成具有插入语性质的定语从句,用以对主句作附加说明,或表示说话者的态度与看法,常可译为"正如……那样""像……",但也可灵活表达。

As has been proved, the mass of the proton is 1,840 times as large as the mass of the electron.

已经证实,质子的质量是电子的 1,840 倍。

Almost all metals are good conductors of electricity, as has been stated above.

如上所述,几乎所有金属都是电的良导体。

Unit Seven Life Science (II)

As mentioned previously, some computers can perform over a billion computations a second.

前面已经提到，有些计算机每秒能运算十亿次以上。

As is shown in the table, the US is the major copper-producing country.

如图所示，美国是主要产铜国。

例60

Air is attracted by the earth, as is every other substance.

空气同其他一切物体一样都受到地球的引力。

科技英语中，由 as 引导的，具有插入语性质的非限制性定语从句有很多，有的已经形成了固定的表达，如：

as described above 如上所述

as explained before 如前面的解释

as already explained 如已解释过的那样

as often happens 正如经常发生的那样

as indicated in... 如在……所表示（指出）的那样

as already mentioned 正如已讲过（提过）的那样

as seen 显然

as shown in... 如……所示

as stated above 如上所述

as pointed out many times 正如多次指出

as is known to all 众所周知，大家都知道的

as is often the case in... 和……的情况一样

as is often said 如通常所说

as might have been expected 如原来所预料的那样

as will be seen later 如将会看到的那样

▶ 2.5.3 由"名词等 + of + which"引导的定语从句

在这种从句中,该名词是从句中的主语,而 of which 则是该名词的定语,翻译时,一般都把 of which 译成"其""它的"或"……的"。整个从句译成一个独立的分句,按原文顺序放在主句之后。

There are 107 known elements, most of which are metals.

已知元素有 107 种,其中大部分是金属。

A laser beam is a coherent light, the waves of which are in step with each other.

激光是一种相干光,其光波是彼此同步的。

This kind of chemicals, the evaporating temperature of which is low, must be kept at a temperature below zero Centigrade.

因为这种化学药品的蒸发温度很低,故其保存温度应在零摄氏度以下。

例64

There are different kinds of solvents, the most important of which is water.

溶剂有多种,而最重要的一种是水。

2·6 定语从句的其他译法

英语中有一些定语从句和主句的关系比较复杂,且表达方式与汉语差别较大,因而需要用其他方法译出。

▶ 2.6.1 转换成同位语从句

有些定语从句和先行词的关系根本不是修饰和限定的关系,这些定语从句阐述或补充说明先行词的具体内容，在逻辑上和先行词处于同等地位。这种定语从句往往可以作为先行词的同位语译出。

例65

The speed of sound in air at ordinary temperature is about 1,100 feet per second, which is about one mile in five second or about 700 miles per hour.

在常温下，声音在空气中的传播速度为每秒 1,100 英尺，即每 5 秒钟约 1 英里或每小时 700 英里。

例66

Continuous cropping is a method of farming in which fields are not given a fallow period between crops.

连作是一种耕作法，即在两次收获之间，土地没有休耕期。

例67

In 1905, which was many years before other scientists really understood a great deal about atomic energy, Einstein declared this theory.

1905 年，即在其他科学家真正对原子能有所了解之前许多年，爱因斯坦就宣布了这一理论。

▶ 2.6.2 转换成主语从句

有些定语从句，尤其是由 why 引导的定语从句，虽然在形式上是从句，但实际上却表达了句子的主要内容。为了突出其作用并便于汉语表达，常可将其转换成主语从句的形式译出。

例68

There appear to be at least three reasons why colloidal particles do not settle out.

为什么胶体粒子不会沉淀，至少有三个原因。

That is the reason why the work put out is always less than that put in.
输出功之所以总小于输入功,原因就在于此。

▶ 2.6.3 和主句合译成简单句

在本单元的开头曾提到,有时可将定语从句逆序译为定语,和主句合译成简单句,这主要指一些结构比较简单的定语从句。这种定语从句和先行词的关系非常密切,在意义上仅相当于全句的一个简单成分。对这种定语从句,翻译时可将其和先行词合并成一个主谓结构,把整个句子译成一个简单句,这种译法在 there be 引导的存在句中特别适用。

A compound is a substance which is composed of the atoms of two or more different elements.
化合物这种物质是由两种或更多种不同元素的原子构成的。

An insulator is a substance which electric current cannot flow through easily.
电流不易通过绝缘体这种物质。

There are also millions of tiny living things that float in the sea.
还有无数微小生物漂浮在海上。

There are some engine rooms on sea-going ships where all the devices are automatically-controlled.
远洋轮船上有些机舱设备是自动控制的。

Unit Seven　Life Science (II)

There are a number of factors that can affect the magnitude of resistance.

有许多因素影响电阻的大小。

2.6.4 加括号译出

当采用上述几种方法无法清晰明了地将定语从句的意思传达出来时，如果硬译，就会使语气不连贯，意义不集中，甚至会冲淡原文的重要意义。此时，如能采用加括号的方法，把定语从句译成注释文字，则可避免句子冗长的弊病，将原文的精髓表达出来。

Some materials, such as cotton, which is often used as insulation, are liable to absorb moisture, and this will adversely affect their insulating properties.

有些物质（如常用作绝缘体的棉花）容易吸水，这对其绝缘性能也有不利的影响。

例76

The research found that the giraffe's jugular vein, which carries blood from the head back to the heart, has lots of one-way valves in it.

研究人员发现，长颈鹿的颈静脉（其作用是把血液从头部送回心脏）中有许多单向瓣膜。

2.6.5 缩译为单词定语

有时，为了使译文简明、生动，可将一些结构简单的定语从句缩译为一个单词，放在先行词之前作定语。这和前面提到的将从句逆序译为前置定语不同，前面提到的方法是把从句译成汉语的主谓结构、动宾结构、偏正结构或动补结构等，全句不只有一个主谓结构；而缩译是把从句译成单个名词或形容词形式的定语，全句只有一个主谓结构。

The human body has some kind of action in itself with which it fights infections.

人体本身具有某种抗感染的能力。

Scientists are able to draw from these germs a substance which is a germ destroyer.

科学家能从这些细菌中提取一种杀菌的物质。

Chlorine is a gas which has a nasty smell.

氯气是一种难闻的气体。

Quiz

1. Translate the following sentences into Chinese. While translating, pay more attention to the italicized words.

    (1) Matter has certain features or properties *that enable us to recognize it easily*.

    (2) Molecules have perfect elasticity, *in consequence of which they undergo no loss of energy after a collision*.

    (3) The quantity of heat possessed by a body depends on several factors, *of which temperature is one only*.

    (4) A typical example of this is the ordinary thermometer, *with which we are well acquainted in our daily life*.

    (5) They worked out a method *by which production has now been rapidly increased*.

    (6) The value of any agent used to do work depends upon the amount of work *it is able to do in a certain time*.

Unit Seven Life Science (II)

(7) This factory produced machine tools **to which precision instruments were attached**.

(8) The sort of lubricant **which we use** depends largely on the running speed of the bearing.

(9) Nowadays it is understood that a diet **which contains nothing harmful** may result in serious disease if certain important elements are missing.

Part B Reading

It's Not "All in the Genes"

It is no surprise that virtually every list that appeared of the most influential people of the 20th century included James Watson and Francis Crick[1], right up there alongside Churchill, Gandhi and Einstein. In **discerning** the double-helical nature of DNA, Watson and Crick **paved the way for** understanding the molecular biology of the gene, the dominant scientific accomplishment of the postwar era. **Sequencing** the human genome will represent a closure of sorts for the revolution wrought by those two geniuses.

At the same time, it's also not surprising that many people get nervous at the prospects of that scientific **milestone**. It will no doubt be a revolution, but there are some scary *Brave New World*[2] **overtones** that raise fundamental questions about how we will think about ourselves. Will it mean that our behaviors, thoughts and emotions are merely the sum of our genes, and scientists can use a genetic roadmap to calculate just what that sum is? Who are we then, and what will happen to our cherished senses of individuality and **free will**? Will knowing our genetic code mean we will know our **irrevocable fates**?

I don't share that fear, and let me explain why. At the **crux** of the anxiety is the notion of the Primacy of Genes. This is the idea that if you want to explain some complex problems in biology (like why some particular bird **migrates** south for the winter, or why a particular person becomes **schizophrenic**), the answer lies in understanding the **building blocks** that make up those phenomena—and that those building blocks are ultimately genes. In this **deterministic** view, the proteins **unleashed** by genes "cause" or "control" behavior. Have the wrong version of a gene and, bam, you're guaranteed something awful, like being **pathologically** aggressive, or

Unit Seven Life Science (II)

having schizophrenia. Everything is **preordained** from **conception**.

Yet hardly any genes actually work this way. Indeed, genes and environment interact: nurture reinforces nature. For example, research indicates that "having the gene for schizophrenia" means there is a 50 percent risk you'll develop the disease, rather than absolute certainty. The disease occurs only when you have a combination of schizophrenia-prone genes and schizophrenia-inducing experiences. A particular gene can have a different effect, depending on the environment. There is genetic **vulnerability**, but not **inevitability**.

The Primacy of Genes also assumes that genes act **on their own**. How do they know when to turn on and off the synthesis of particular proteins? If you view genes as **autonomous**, the answer is that they just know. No one tells a gene what to do; instead, the buck[3] starts and stops there.

However, that view is far from accurate too. Within the **staggeringly** long sequences of DNA, it turns out that only a tiny percentage of letters actually form the words that constitute genes and serve as code for proteins. More than 95 percent of DNA, instead, is "non-coding". Much of DNA simply constitutes on and off switches for regulating the activity of genes. It's like you have a 100-page book, and 95 of the pages are instructions and advice for reading the other five pages. Thus, genes don't independently determine when proteins are synthesized. They follow instructions originating somewhere else.

What regulates those switches? In some instances, chemical messengers from other parts of the cell. In other cases, messengers from other cells in the body (this is the way many hormones work). And, critically, in still other cases, genes are turned on or off by environmental factors. As a crude example, some **carcinogens** work by getting into cells, binding to one of those DNA switches and turning on genes that cause the uncontrolled growth that constitutes cancer. Or a mother rat **licking** and **grooming** her infant will initiate **a cascade of** events that eventually turn on genes related to growth in that child. Or the smell of a female **in heat** will **activate** genes

157

in certain male **primates** related to reproduction. Or a miserably stressful day of final exams will activate genes in a typical college student that will suppress the immune system, often leading to a cold or worse.

You can't **dissociate** genes **from** the environment that turns genes on and off. And you can't dissociate the effects of genes from the environment in which proteins exert their effects. The study of genetics will never be so all **encompassing** as to **gobble up** every subject from medicine to sociology. Instead, the more science learns about genes, the more we will learn about the importance of the environment. That goes for real life, too: genes are essential but not the whole story.

New Words and Expressions

discern 辨出，识别
pave the way for... 为……铺平道路
sequence 排序
milestone 里程碑
overtone 含意；弦外之音
free will 自由意志
irrevocable 注定的
fate 命运
crux 症结；难关，难题
migrate 迁徙
schizophrenic 精神分裂症的
building block 基本因素
deterministic 宿命论的，决定的
unleash 释放
pathologically 病理地
preordain 预先注定

conception 观念；怀孕
vulnerability 脆弱性
inevitability 必然性
on one's own 独自地
autonomous 自发作用的；自治的
staggeringly 令人惊愕地
carcinogen 致癌物质
lick 舔
groom 梳理（毛发）；整理
a cascade of 一系列
in heat 发情期
activate 激活
primate 灵长类动物
dissociate...from 将……分开
encompassing 包罗万象的
gobble up 狼吞虎咽

Unit Seven Life Science (II)

📝 Notes

1. James Watson and Francis Crick 詹姆斯·沃森和弗朗西斯·克里克。

   1928年4月6日，沃森出生于美国芝加哥。他16岁就从芝加哥大学毕业，获动物学理学学士学位，并在生物学方面开始显露才华。22岁时沃森来到英国剑桥大学的卡文迪什实验室，结识了早先已在这里工作的克里克，从此开始了两人传奇般的合作生涯。

   克里克于1916年6月8日出生于英格兰的北安普敦，21岁从伦敦大学毕业。第二次世界大战结束后，他来到剑桥大学的卡文迪什实验室，克里克是深受薛定谔的《生命是什么？》一书的影响，从物理学转向研究生物学。

   他们通过对大量X射线衍射材料的分析研究，提出了DNA的双螺旋结构模型，这一研究于1953年4月25日在英国《发现》杂志正式发表，并由此建立了遗传密码和模板学说。

   之后，科学家们围绕DNA的结构和作用，继续开展研究，取得了一系列重大进展，并于1961年成功破译了遗传密码，以无可辩驳的科学依据证实了DNA双螺旋结构的正确性，从而使沃森、克里克同威尔金斯一道于1962年获得诺贝尔医学生理学奖。

2. Brave New World《勇敢的新世界》是英国作家奥尔德斯·赫胥黎（Aldous Huxley）的一部科幻小说，描述了一个人被基因控制的恐怖世界。

3. 句中 buck 指基因合成蛋白质的工作机制。

📝 Exercises

1. Find out the English equivalents of the following Chinese terms from the passage.

   （1）DNA 双螺旋性　　　　　（2）基因分子生物学
   （3）人类基因组　　　　　　（4）基因图谱
   （5）遗传密码　　　　　　　（6）基因至上
   （7）易患精神分裂症的基因　（8）诱发精神分裂症的经历

2. Translate the following sentences into Chinese.

(1) It is no surprise that virtually every list that appeared of the most influential people of the 20th century included James Watson and Francis Crick, right up there alongside Churchill, Gandhi and Einstein.

(2) At the same time, it's also not surprising that many people get nervous at the prospects of that scientific milestone. It will no doubt be a revolution, but there are some scary *Brave New World* overtones that raise fundamental questions about how we will think about ourselves.

(3) Have the wrong version of a gene and, bam, you're guaranteed something awful, like being pathologically aggressive, or having schizophrenia. Everything is preordained from conception.

(4) Yet hardly any genes actually work this way. Indeed, genes and environment interact: nurture reinforces nature.

(5) If you view genes as autonomous, the answer is that they just know. No one tells a gene what to do; instead, the buck starts and stops there.

(6) Or a miserably stressful day of final exams will activate genes in a typical college student that will suppress the immune system, often leading to a cold or worse.

(7) The study of genetics will never be so all encompassing as to gobble up every subject from medicine to sociology. Instead, the more science learns about genes, the more we will learn about the importance of the environment. That goes for real life, too: genes are essential but not the whole story.

Part C Extended Reading

Genetically Modified Foods

Genetically modified foods have already arrived on American dinner tables. But are they safe?

While it can't yet be said that every mouthful of food has been changed through genetic engineering, it is likely that almost every American has had a mouthful of engineering food.

Take the **soybean**, for example. About 55 percent of last year's crop was genetically engineered in some **fashion**, reports the American Soybean Association. And though few Americans sit down to a plate of soybeans for dinner, these legumes arrive through the back door as an additive to **scores of** foods like mayonnaise, margarine, cooking oils, salad dressings, coffee creamers, beer, cereals, candy, and shortenings.

America is **on the verge of** a second green revolution. The first tripled world food output over the course of a mere three decades during the late 20th century. Scientists **boosted** crop yields by **crossbreeding** related plants that add desirable traits. Then farmers added fertilizer, pesticides, and irrigation in order to make the high-yield crops thrive.

Now that those **gains** are **leveling off**, scientists are looking to biotechnology to increase food production even more. This time, instead of breeding plants with their closest relatives, scientists are inserting genes. For example, genes from **flounders** can help ordinary plants like tomatoes and strawberries fight the cold. Researchers are also inserting bacterial genes into corn and soy-bean plants to better protect them from insects or render them **immune to** certain herbicides.

The technology has enormous potential. "The application of biotechnology is going to produce a set of possibilities that we simply cannot conceive of, even in our most imaginative flights of fancy."

For farmers, crops engineered with genes that resist cold, drought, or other **adverse** weather conditions can boost crop yields with less money and effort. For consumers, that could mean cheaper food. Such crops would also help feed and better nourish people in developing nations, such as **drought**-ridden Africa.

And **splicing** genes into crops such as corn to make them secrete their own pesticide may be better for the environment; farmers may not need to spray their fields with as much pesticide. For the future, expect foods that can help prevent cancer, fruits and vegetables that deliver vaccines, and crops that even make their own fertilizer.

But as these crops begin **infiltrating** our food supply, environmental and consumer groups have begun to question whether potential risks to the environment and human health have been adequately studied. Last year, for example, biologists for the first time found evidence suggesting that planting genetically modified corn in open fields may kill butterflies who feed on the corn's **pollen**. Scientists are also questioning whether foods with a gene inserted to improve one area of their performance could prove **detrimental** in another: For one thing, the genes might cause allergic reactions in people who never had a reaction to that food before.

In the mid-1990s, for example, Des Moines, Iowa-based Pioneer Hi-Bred International sought to boost the quality of soybeans by introducing a Brazil nut gene. But after tests found that the modified soybean gave people allergic to Brazil nuts **hives**, the company **scrapped** the project. Allergies to nuts are common, and so testing the genetically modified soybean for a reaction **made sense**.

But engineering a crop takes a lot of **trial and error**. When a scientist inserts a new gene into a crop, he'll get several different types of seed. Some will have the gene inserted in the wrong place, some will have no gene at all. Eventually, the researcher will get the seed he wants. But what if the gene also subtly inserted itself somewhere else in the plant's DNA?

Unit Seven Life Science (II)

New Words and Expressions

soybean 大豆
fashion 方式
scores of 大量的
on the verge of 接近于
boost 推进；提高；支持
crossbreed 杂交；杂种
gain 收获
level off 平稳；达到稳定
flounder 比目鱼
immune to... 对……有免疫力

adverse 不利的
drought-ridden 干旱的
splice 拼接；连接
infiltrate 侵入
pollen 花粉
detrimental 有害的，不利的
hives 荨麻疹
scrap 废弃（计划、约定等）
make sense 有意义
trial and error 反复试验，不断摸索

Exercises

1. Find out the English equivalents of the following Chinese terms from the passage.

    （1）豆科植物 （2）添加剂
    （3）蛋黄酱 （4）人造黄油
    （5）食用油 （6）沙拉酱
    （7）咖啡伴侣 （8）谷类食品
    （9）起酥油 （10）化肥
    （11）杀虫剂 （12）灌溉
    （13）高产作物 （14）除草剂
    （15）过敏反应

2. Translate the following sentences into Chinese.

    (1) While it can't yet be said that every mouthful of food has been changed through genetic engineering, it is likely that almost every American has had a mouthful of engineering food.

    (2) This time, instead of breeding plants with their closest relatives, scientists are inserting genes.

(3) The technology has enormous potential. "The application of biotechnology is going to produce a set of possibilities that we simply cannot conceive of, even in our most imaginative flights of fancy."

(4) Scientists are also questioning whether foods with a gene inserted to improve one area of their performance could prove detrimental in another: For one thing, the genes might cause allergic reactions in people who never had a reaction to that food before.

Unit Eight
Aeronautics

Part A Lecture

长句的翻译

由于英语和汉语的语法和表达习惯不同，英语中大量错综复杂的长句是翻译中的难点。英语中有大量语法功能很强的介词、连词，有丰富的词性变化，有各种短语和不同从句，还有许多语法手段，经常主句中有从句，短语中有短语。这使得英语句子层次繁杂，像一棵枝繁叶茂的大树，盘根错节。在书面语中，英语为了表达严谨，多用长句。在科技英语中，为了说理严谨，逻辑紧密，描述准确，更是有大量长句出现。因此，我们有必要掌握一定的长句翻译技巧，才能正确翻译。

汉语句子在表达复杂的概念时多用短句，少用关联词语，以语序、意合表达其中的层次关系，且句子简短、节奏明快。在翻译科技英语长句时，如果按照英语句子结构翻译成汉语，则必然生硬难懂，甚至使人不明所以。

英文长句汉译，一般要拆成汉语短句，按照汉语的表达习惯和逻辑层次，重新排列顺序。组织成内容准确、逻辑分明、重点突出、通顺正确的译文。

翻译时首先要通读全句，分清句法结构。先分析句子是简单句、并列句还是复合句。英语句子除了并列句外，一个句子只有一个主句，其他起修饰作用的词、短语、从句都是次要成分。要正确划分主要成分和次要成分的范围和层次关系，语法关系不清会导致理解乃至翻译上的错误。要注意长句的连接手段，如连词、介词、非谓语形式（不定式短语、分词短语、动名词短语）、各种从句（主语从句、宾语从句、同位语从句、定语从句、状语从句）。

理解语法关系之后要领会句子的要旨和逻辑关系，分清重心，突出重点。英语句子习惯先传递主要信息，加上各种连接手段，再补充各种次要信息。英语的重心也可后置，应灵活对待。而汉语句子多用语序表达逻辑层次，一般是由小到大、先因后果、重心在后。

在译文表达阶段，先将英语长句中每个短句译成汉语，然后按照汉语的语序，重新排列组合，以正确表达原意。在此阶段语序不必与原句一致。最后加工润色。

Unit Eight Aeronautics

英语长句一般有以下译法：分译法、顺译法、倒译法和变序法。下面将对这些译法作一介绍。

1 分译法

科技英语中长句居多，按原句照译会使汉语句子过于庞杂臃肿，故而可以把英语的某些句子成分单独译为一个汉语句子。

例1

Each cylinder therefore is encased in a water jacket, which forms part of a circuit through which water is pumped continuously, and cooled by means of air drawn in from the outside atmosphere by large rotary fans, worked off the main crankshaft, or, in the larger diesel electric locomotives, by auxiliary motors.

原译：因而每个气缸都用一个形成循环回路的一部分，由水泵驱使水在回路中不停地流动，并由外部鼓进的空气来使水冷却（鼓风用的大型旋转风扇是由主曲轴带动的，而在大型电传动内燃机车上，则由辅助电机带动）的水套围着。

分析：本句主句为 Each cylinder therefore is encased in a water jacket，其余是由 which 引导的定语从句 which forms part of a circuit through which water is pumped continuously, and cooled by means of air drawn in from the outside atmosphere by large rotary fans, worked off the main crankshaft, or, in the larger diesel electric locomotives, by auxiliary motors。初译的汉语译文过于复杂，定语过长，应改译。

改译：因而每个气缸都用一个水套围着，水套形成循环回路的一部分，由水泵驱使水在回路中不停地流动，并用由外部鼓进的空气来使水冷却。鼓风用的大型旋转风扇是由主曲轴带动的，而在大型电传动内燃机车上，则由辅助电机带动。

Half-lives of different radioactive elements vary from as much as 900 million years for one form of uranium, to a small fraction of a second for one form of polonium.

不同的放射性元素，其半衰期也不同。有一种铀的半衰期长达9亿年，但有一种钋的半衰期却短到几分之一秒。

167

An infra-red system could be useable in both anti-air and anti-ship engagements, but its inherent disadvantages related to some dependence on optical visibility and to sensibility to interference from natural or man-made sources make it less attractive in the surface than in the air role.

红外系统既可用于对空作战，也可用于对舰作战，但是它有两个固有的缺点：一是对光学能见度有一定的依赖性；二是易受天然或人造干扰源的干扰。因此，该系统的对空作用较之对舰作用更为诱人。

From what is stated above, it is learned that the sun's heat can pass through the empty space between the sun and the atmosphere that surrounds the earth, and that most of the heat is dispersed through the atmosphere and lost, which is really what happens in the practical case, but to what extent it is lost has not been found out.

由上述可知，太阳的热量可以穿过太阳与地球大气层之间的真空，而大多热量在通过大气层时都扩散和消耗了。实际发生的情况正是如此，但是热量的损失究竟达到什么程度，目前尚未弄清。

2 顺译法

当英语长句的表达顺序与汉语的习惯表达顺序一致时，可不改变原文语序和语法结构，译成汉语。但值得一提的是，并不是每一个词都按照原句顺序翻译，事实上一一对应是极少见的。试看下面例句。

During the high energy period of a physical biorhythm we are more resistant to illness, better coordinated and more energetic; during the low energy period we are less resistant to illness, less well coordinated and tired more easily.

分析：本句原文为一个并列句，各分句表达顺序在汉语中也很顺畅，故按原文顺序。个别词语顺序应稍做调整。

译文：在身体生物节奏的高能期，我们有较强的抗病能力，身体各部分更

Unit Eight　Aeronautics

协调自如，精力更旺盛；在低能期，我们的抗病能力减弱，身体各部分不太协调一致，而且容易感到疲劳。

　　Many man-made substances are replacing certain natural materials because either the quantity of the natural products cannot meet our ever-increasing requirement, or more often, because the physical properties of the synthetic substance, which is the common name for man-made materials, have been chosen, and even emphasized so that it would be of the greatest use in the field which it is to be applied.

　　许多人造材料正在取代某些天然材料，这或者因为天然产品的数量不能满足人类不断增长的需要，或者更多的因为合成物（这是各种人造材料的统称）的物理性能被选中，甚至受到极大的重视，因而使得它会在准备加以采用的领域获得最大程度的应用。

　　Vibration in machines can be thought of as a combination of non-stationary periodic functions generated by a variety of imbalanced forces or disturbances, each of which has a characteristic repetition frequency.

　　可以把机器的振动视作一些非稳态周期性作用的综合反应，这些作用是由各种不平衡力或扰动引起的，每一种不平衡力或扰动都有一个特殊的重复频率。

　　Chemists study the structure of food, timber, metals, drugs, petroleum and everything else we use to find out how the atoms are arranged in molecules, what shapes the molecules have, what forces make the molecules arrange themselves, into crystals, and how these crystals arrange themselves into useful substances.

　　化学家们研究食物、木材、金属、药品、石油以及其他一切我们所用的东西的结构，以期发现其原子在分子中是如何排列的，分子是什么样的，是什么力量使那些分子排列成为晶体，而晶体又是如何排列成有用的物质的。

3 倒译法

有些英语长句的表达顺序与汉语习惯不同，甚至相反。如主句后有表示原因、时间、地点、方式等的状语从句，或有后置的定语从句、主语从句等。翻译时须逆着原文顺序，把原文后面的内容放在译文的前面。

The technical possibility could well exist, therefore, of nationwide integrated transmission network of high capacity, controlled by computers, inter-connected globally by satellite and submarine cable, providing speedy and reliable communications throughout the world.

原译：这种可能性从技术上来讲是完全存在的，建立全国统一的大容量的通信网络，由计算机进行控制，通过卫星和海底电缆为全球提供快速可靠的通讯服务。

分析：本句主句为 The technical possibility could well exist，其余是很长的 of 介词短语做 possibility 的定语。原译的汉语译文照搬原文顺序，层次不清。本句语义重心为：这种可能性从技术上来讲是完全存在的，在汉语中应放在句尾，加以总结。

改译：建立全国统一的大容量的通信网络，由计算机进行控制，通过卫星和海底电缆为全球提供快速可靠的通讯服务，这种可能性从技术上来讲是完全存在的。

Being able to receive information from any of a large number of separate places, carry out the necessary calculations and give the answer or order to one or more of the same number of places scattered around a plant in a minute or two, or even in a few seconds, computers are ideal for automatic control in process industry.

分析：本句是一个简单句，"Being...in a few seconds..."为现在分词短语作状语，主句很短。

译文：计算机用于程序生产的自动控制是理想的。它能在一两分钟，甚至几秒钟内，从分散在工厂周围的许多点得到信息，进行必要的运算，给其中发出信息的一个或者更多的点以回答，或发出指令。

Unit Eight　Aeronautics

Various machine parts can be washed very clean and will be as clean as new ones when they are treated by ultrasonic waves, no matter how dirty and irregularly shaped they maybe.

各种机器零件，无论多么脏，不管形状多么不规则，当用超声波处理后，都可以清洗得非常干净，甚至像新零件一样。

The increasing speed of scientific development will be obvious if one considers that television, space craft, and nuclear-powered ships, which are taken for granted now, would have seemed fantastic to people whose lives ended as recently as the 1920s.

电视机、宇宙飞船和核动力舰艇对今天的人们来说已不是什么新鲜事物，但对于近在20世纪20年代去世的人来说，还似乎是不可思议的东西。考虑到这一点，科学发展的速度之快也就显而易见了。

Physical science had emerged from its incipient stages **with** definite methods of hypothesis and induction and of observation and deduction, **with** the clear aim to discover the laws by which phenomena are connected with each other, and **with** a fund of analytical process of investigation.

分析：本句是带有一个定语从句"by which...with each other"的主从复合句。主句中有三个作状语的并列介词短语（with + 宾语）。

译文：自然科学从其形成的最初阶段开始，就有关于假说和归纳、观察和推论的特定方法，有明确的目标，即要发现把各个现象联系在一起的规律，还有在研究过程中所采用的许多分析方法。

4　变序法

有的英语长句语法成分繁多、层次复杂，顺着或逆着原文的顺序翻译成汉语皆不够通顺或层次错乱，以致难以理解。在这种情况下，须打乱原文顺序，按照汉语表达习惯正确表达原文的层次关系和语义关系。这样虽然与原文语序不同，但能准确达意。

Since natural water contains dissolved oxygen, fishes, which also need oxygen though they live in water, will die if hot days persist so long that the dissolved oxygen is evaporated.

原译：因为天然水含有溶解的氧，鱼却也需要氧气，虽然生活在水中，鱼就会死去，如果天热时间过长，以至于水中的溶解氧都蒸发了。

分析：本句主句为 ... fishes will die...，句中有 since 和 so...that 引导的两个状语从句。原译的汉语译文逻辑不清、层次杂乱，应调整顺序改译。

改译：鱼虽然生活在水中，却也需要氧气，而天然水含有溶解的氧。因此，如果天热时间过长，水中的溶解氧都蒸发了，鱼就会死去。

Atomic nuclei consist of combinations of protons, or positively-charged particles, and neutrons, or uncharged particles.

原子核是由质子与中子相结合而构成，质子也叫带正电荷的粒子，中子也叫不带电荷的粒子。

The final part of the paper highlights the areas for potential application of thermo-mechanical treatments and emphasizes the need for information to facilitate design of suitable forming equipment for exploiting the potential of thermo-mechanical treatments over a range of temperatures as a means of producing various product forms with enhanced property combinations.

文章的最后部分着重论述形变热处理可能应用的领域，并强调指出：为了便于设计适宜的成形设备，作为生产具有优异综合性能的不同形状产品的工具，需要获得有关的知识，而这种成形设备可用来发挥在一系列温度下形变热处理的潜力。

One of the significant fringe benefits of the remarkably small size of integrated circuit systems is that effective external nuclear radiation shields are now feasible, where weight or cost ruled them out before.

集成电路装置的尺寸特别小，这就又增加了一个很大的优点：可以在外界给它提供有效的核辐射屏蔽；这在以前，由于重量或价格的限制是不可能做到的。

Quiz

1. **Translate the following sentences into Chinese.**

    (1) The most frequent forms of the arthritis are osteoarthritis, which is increasingly common with age, and rheumatoid arthritis, which can strike at any age, even during infancy.

    (2) With others, notably lead, mercury and tin, there was a temperature, different for each one but well below zero, at which the resistance dropped to nothing at all.

    (3) They had to find the explanation of a growing series of results which not only seemed contrary to nature as man understand it, but which did not at first fit in too well with each other.

    (4) The physical metallurgist, who has been relied upon by the industrialist to solve immediate problems even though it may have been at the expense of his fundamental work on the understanding of the overall behavior of metals and alloys, is now tempted to desert to the ranks of the metal physicists, although his training equips him ill for the new occupation.

    (5) Another application is in the process known as electro-plating, in which a thin surface of some metal such as chromium or tin is deposited on a metallic base so that it adheres firmly to the base.

    (6) With the present rate of growth of ideas and plans of both physicists and engineers, it is no longer possible to allow

the development of the full potentialities of the new metals to evolve over a period of about fifty years, as was the case with aluminum in the period between 1890 and 1949.

(7) However, the proponents of the above would perhaps be surprised to learn that the German Marine have concluded that the best weapon for engaging heavily defensed naval targets are good old iron bombs.

(8) Through scientific practice we've come to realize that matter can neither come from naught nor can it be reduced to naught; the total mass of reagents before reaction is constantly equivalent to that of resultants after reaction—the law of conservation of matter.

(9) The light cruiser and frigate, accompanied by an oiler, were sighted heading south, the spokesman said, apparently on their way to join the other ships in the South China Sea area.

(10) Computer language may range from detailed low level close to that immediately understood by the particular computer, to the sophisticated high level which can be rendered automatically acceptable to a wide range of computers.

(11) The secret of the moon remained unveiled until the latter half of the twentieth century due to the lack of lunar space carrier, for a series of questions concerning fuel, material, safe landing, propelling mechanism and particularly electronic computation etc. were too intricate to solve under the technical conditions early in this century, which have since been undergoing profound change and greatly improving.

Unit Eight   Aeronautics

(12) There are rivers that at some seasons of the year have very much more water than at others, rivers that are made to generate vast quantities of electricity by their power, and rivers that carry great volumes of traffic.

(13) The objective of this document is to provide the Conference with a ready reference to earlier decisions so that discussions on current issues could be facilitated without undue time being spent on issues already dealt with.

Part B Reading

The Plane Makers

There are two main things that make aircraft engineering difficult: the need to make every **component** as reliable as possible and the need to build everything as light as possible. The fact that an aeroplane is up in the air and cannot stop if anything goes wrong makes it perhaps a matter of life or death that its performance is absolutely dependable.[1]

Given a certain power of engine, and consequently a certain fuel consumption, there is a practical limit to the total weight of aircraft that can be made to fly.[2] Out of that weight as much as possible is wanted for fuel, radio **navigational** instruments, passengers seats, or **freight** room, and, of course, the passengers or freight themselves. So the structure of the aircraft has to be as small and light as safety and efficiency will allow.[3] The designer must calculate the normal load that each part will bear. This specialist is called the "**stress man**". He **takes account of** any unusual stress that may be put on the part as a **precaution** against errors in manufacture, accidental damage, etc.[4]

The stress man's calculations go to the designer of the part, and he must make it as strong as the stress man says is necessary.[5] One or two samples are always tested to prove that they are as strong as the designer intended. Each separate part is tested, then a whole assembly—for example, a complete wing, and finally the whole aeroplane. When a new type of aeroplane is being made, normally only one of the first three made will be flown. Two will be destroyed on the ground in structural tests. The third one will be tested in the air.

Two kinds of ground strength tests are carried out. The first is to find the **resistance** to loading of the wing, tail, etc. until they reach their maximum load and collapse. The other test is for fatigue strength.

Relatively small loads are applied thousands of times. Each may be well under what the structure could stand as a single load, but many repetitions can result in **collapse**.[6]

When a plane has passed all the tests it can get a government certificate of **airworthiness**, without which it is illegal to fly, except for test flying.[7]

Making the working parts reliable is as difficult as making the structure strong enough. The flying controls, the electrical equipment, the fire precautions, etc. must not only be light in weight, but must be below freezing point and in the hot air of an airfield in the tropics.

To solve all these problems the aircraft industry has a large number of research workers, with **elaborate** laboratories and test houses, and new materials to give the best strength **in relation to** weight are constantly being tested.

New Words and Expressions

component 部件
given 假设；作为前提
navigational 航行的；航行用的
freight 货物
stress 压力；应力
take account of 考虑
precaution(against...)（对……的）预防措施

resistance 抵抗力
collapse 崩溃；倒塌
airworthiness 适航性；飞行技能
elaborate 精细的，精心制作的
in relation to 有关

Notes

1. 句中 an aeroplane is up in the air and cannot stop if anything goes wrong 是 the fact 的同位语，意为"如果发生故障，飞机在空中飞行，并且不能停飞，这一事实……"。句中 it 为形式宾语，真正宾语是 its performance is absolutely dependable。

2. 句中 that can be made to fly 修饰 the total weight of aircraft，意为"飞机可以升空飞行的总重量"。

3. 全句译为：因此，在安全和效率允许的情况下，飞机的结构必须尽可能地做到小且轻。

4. 句中 that may be put on the part 修饰 any unusual stress；as 意为"作为"。全句译为：他要考虑可能加在部件上的任何异常应力，作为预防制造中可能产生的误差、意外损坏等的措施。

5. 句中 calculations 翻译成"计算结果"。全句译为：应力计算专家将计算结果交给部件的设计人员，部件设计人员必须使部件达到应力计算专家所要求的强度。

6. 全句译为：作为单独的一个载荷来说，每一次（载荷量）都可能远远低于结构所能承受的限度，但是多次重复以后即可导致飞机毁坏。

7. 句中 which 指的是 certificate。全句译为：飞机通过这一切试验之后，就可以获得政府颁发的适航证。如果没有这种证书，除了试飞之外，一切飞行都是不合法的。

Exercises

1. Find out the English equivalents of the following Chinese terms from the passage.

（1）飞机工程　　　　　　　　（2）一个生死攸关的问题
（3）发动机功率　　　　　　　（4）耗油率
（5）无线电导航仪器　　　　　（6）旅客座位
（7）货舱　　　　　　　　　　（8）正常载荷
（9）应力计算专家　　　　　　（10）意外损坏
（11）地面强度试验　　　　　（12）疲劳强度
（13）适航证　　　　　　　　（14）飞行控制机构
（15）电力设备　　　　　　　（16）防火设施
（17）冰点以下温度

2. Translate the following sentences into Chinese.

(1) There are two main things that make aircraft engineering difficult: the need to make every component as reliable as possible and the need to build everything as light as possible. The fact that an aeroplane is up in the air and cannot stop if anything goes wrong makes it perhaps a matter of life or death that its performance is absolutely dependable.

(2) Given a certain power of engine, and consequently a certain fuel consumption, there is a practical limit to the total weight of aircraft that can be made to fly.

(3) He takes account of any unusual stress that may be put on the part as a precaution against errors in manufacture, accidental damage, etc.

(4) Two kinds of ground strength tests are carried out. The first is to find the resistance to loading of the wing, tail, etc. until they reach their maximum load and collapse.

(5) Each may be well under what the structure could stand as a single load, but many repetitions can result in collapse.

Part C Extended Reading

How Aircraft Are Built

The stage when the construction of the new aircraft begins is known as "cutting metal"; this is a most **apt** term, as will be seen later on. Just as the design of an aircraft is a most complicated matter, so too is its manufacture, which involves the use of complex and expensive machinery, and **taxes** to the limit the human capacity for organization and planning. The organization and planning involved is probably something not often considered, yet without this no new airliner would ever take to the air.

One of the reasons why the construction of a new aircraft is such a complex matter is that in order to help to ensure that the aircraft is a definite improvement on its **predecessors**, many of the system components, such as fuel valves, pumps, electrical generators and **switches**, and hydraulic pumps and **jacks**, will be new, and not existing **components** available "off the shelf". The design and development of these **accessary** components, and there are many hundreds of them, are usually carried out by specialists in the various fields. The development of some of the components, such as the electronic units for the auto-pilot and navigational systems is extremely complex, requiring equipment every bit as expensive and complex as any used by the air-frame manufacturer himself. It is thus no easy matter, when the aircraft designer himself probably does not know exactly what he wants, to look ahead and say that such and such a component will take "two years and four months" to produce. Yet, whatever the degree of complexity, each component is required at a specific time to **suit** the overall production program. The program is always **compressed** as much as possible, for once the decision is taken to "go ahead" with a new aircraft, it is usually commercially advantageous to get it built and flying as soon as possible.

Unit Eight   Aeronautics

As far as the construction of the air-frame is concerned, work naturally starts first on those parts that take a long time to **fabricate** and that are required early in the program. Many key attachment **fittings** are machined from **castings**, **extrusions** and **forgings**, and the castings and **billets** are ordered early. Metal cutting and machining usually start on fuselage frames and the wing **spars**.

Construction of the wings and fuselage begins with the **jigging**, or supporting, of the main spars and **ribs** and fuselage frames, upon which the remainder of the structure is built.

The traditional method of making fuselage and wings is from metal sheet, **stiffened** by **riveted stringers**, and riveted and **bolted** to the fuselage frames and wing spars.

Increase in the performance and size of aircraft have **necessitated** the development of new manufacturing techniques, of which the machined skin **panel** is one. During this process the various formers, stringers and stiffeners around access-panel cut-outs, are machined integrally with the skin. The skin itself is often machined to varying thickness, being thinned down where the loads are lowest to save every possible **ounce** in weight.

This method of construction, by eliminating rivet holes, **eases** the problem of **sealing integral** fuel tanks and **pressurized** fuselages. More important still, good fatigue properties can be obtained more readily. This is due to the absence of regions of high localized loading which occur at rivet positions, and also to the fact that large **radii** can be machined at intersection points to "smooth out" load distributions. An integrally machined skin panel can often be made lighter than its fabricated **counterpart** of **equivalent** strength.

There are several advantages in using automatic controls, including the saving of the time that would otherwise be spent in **tedious** manual positioning of the cutter and a great improvement in positional accuracy—to about one tenth of a thousandth of an inch. In addition, cutter positions can be reproduced exactly at any time, no complex and expensive boring jigs are required and if a change of design should occur all that is needed is a new set of punched-card instructions instead of a new jig!

New Words & Expressions

apt 适当的，合适的
tax 使受压力，使负重担
predecessor 前任（人，辈）；上一代
switch 电门；开关
jack 制动器，动作筒
component 部分；部件，元件
accessary 附件，附属品；附属的，附加的
suit 适合于，使适应
compress 压缩
fabricate 制作，制造
fitting 连接件，配件，零件；设备，装置
casting 铸件；铸造
extrusion 压出品，挤压型材
forging 锻；锻件
billet （金属的）坯段，钢坯
spar 梁

jig 钻孔模；夹紧，紧固
rib 翼肋，加强筋
stiffen 加固，加强，增强
rivet （用铆钉）铆；固定
stringer 桁条，长桁
bolt 螺栓；闩上
necessitate 使成为必需，使需要
panel 板，仪表板
ounce 盎司，英两
ease 减轻；放松；使缓和
seal 密封
integral 整体的；完全的
pressurized 加压的，受压的
radii （radius 的复数）半径
counterpart 配对物；互换件；副本
equivalent 相等的，相同的
tedious 令人厌烦的；繁重的

Exercises

1. Find out the English equivalents of the following Chinese terms from the passage.

（1）燃油阀　　　　　　　　　（2）发电机
（3）液压泵　　　　　　　　　（4）自动驾驶仪

(5) 导航系统　　　　　　（6) 在商业上有利可图的
(7) 机身　　　　　　　　（8) 穿孔卡片

2. Translate the following sentences into Chinese.

(1) Just as the design of an aircraft is a most complicated matter, so too is its manufacture, which involves the use of complex and expensive machinery, and taxes to the limit the human capacity for organization and planning.

(2) One of the reasons why the construction of a new aircraft is such a complex matter is that in order to help to ensure that the aircraft is a definite improvement on its predecessors, many of the system components, such as fuel valves, pumps, electrical generators and switches, and hydraulic pumps and jacks, will be new, and not existing components available "off the shelf".

(3) The development of some of the components, such as the electronic units for the auto-pilot and navigational systems is extremely complex, requiring equipment every bit as expensive and complex as any used by the air-frame manufacturer himself.

(4) Construction of the wings and fuselage begins with the jigging, or supporting, of the main spars and ribs and fuselage frames, upon which the remainder of the structure is built.

(5) There are several advantages in using automatic controls, including the saving of the time that would otherwise be spent in tedious manual positioning of the cutter and a great improvement in positional accuracy—to about one tenth of a thousandth of an inch.

Unit Nine
Astronautics

Part A　Lecture

段落的翻译

段落是由句子组成的语义单位，它既是文章的组成部分，又是自成一体、相对独立的整体。因此，段落的翻译以句子翻译为基础，但又不简单地停留在句子翻译层面，而是更多地把段落作为一个整体来考虑，如考虑段落的主题、段落中句子与句子之间的衔接和连贯等问题。

1 英汉段落比较

由于英语与汉语分属不同语系，因而两种语言在句子和段落结构上有所不同。在句子方面，英语句子重形合，注重形式结构的特点，句子与句子之间的连接有一定的规律可循；汉语句子重意合，句子与句子之间的连接较弱，结构上的黏着性不强。在段落结构方面，英语的段落多为演绎式，起首开宗明义，直点主题；段落结构往往是主题句出现在段首，段落中的其他句子一方面以语义与主题句直接关联，另一方面以一些逻辑连接词显现逻辑序列和句际关系。因此，英语段落条理清晰，层次结构清楚，衔接自然流畅，句与句之间有内在的逻辑关系。而汉语的段落多为归纳式，讲究"起、承、转、合"，考虑时间事理顺序，段落主题往往会出现在不同的地方，有时还不止一次地出现，有时是隐含的；同时，句际之间的意义关联可以是隐约的、似断非断的。因此，汉语段落是以某一中心思想统领的形散神聚结构，没有英语中常见的那些连接词。下面试分析英语段落的结构。

(1) *A natural resource like coal can be lost in at least three ways—more than most people realize.* (2) *Firstly,* we can lose a natural resource by using it up or by using it far faster than it can be replenished. (3) Thus we lose coal by burning it. (4) *Secondly,* we can lose a natural resource by letting it be wasted, as when we allow farmland to erode. (5) Coal can be wasted by allowing a mine to become inoperable, or by using inefficient methods

of burning it. (6) **Thirdly,** we can so mismanage the waste products of a natural resource that they pollute or destroy other natural resources. (7) The draining of coal waste into a freshwater river would harm wild life as well as needed supplies of pure water. (8) We might even go a **fourth** step, to say that human labor is also a natural resource, which can be lost by exhaustion, misuse, or no use. (9) Thus the Pogatab Creek flood reduced this resource by putting the inhabitants out of work.

分析：本段落层次结构非常清楚。第一句概括主题，后面的句子列举了造成自然资源浪费的四种原因，每一种原因都用两句话叙述，第一句概括论述，第二句是具体例证。四种原因之间使用逻辑连接词（firstly、secondly、thirdly、thus 等），句与句之间衔接成为一个有机整体。

2 段落翻译的方法

由于英语和汉语段落结构的不同，我们翻译时应先将段落作为一个有机整体进行分析，注意段落的主题以及段落句间的衔接与连贯。

2.1 注意段落的主题

一般而言，段落往往有一个主题，英语段落内的每句话都紧密地围绕主题或主题句展开。因此，翻译也应围绕主题或主题句进行。

▶ 2.1.1 从主题入手——突出主题信息

翻译时，把握主题可以保证整个段落的主题思想和主要信息的正确性，可以使语气连贯，使段内句子有机地联系在一起。

Ultraviolet rays are waves similar to light which are just beyond the violet end of the visible light spectrum. *Ultraviolet rays* are sometimes called invisible light or black light because we cannot see **them** with the human eye. *Ultraviolet rays* have wavelengths between 12 and 390 millimicrons. *The rays* with wavelengths between 300 and 390 millimicrons make people's skin tan. ***They*** also change a chemical in the skin into Vitamin D. ***Those*** between 185 and 300 millimicrons are used for killing harmful bacteria.

译文：**紫外线**是一种与光类似的波，**它们**超越可见光谱的紫色终端。因为我们肉眼看不到**紫外线**，所以**它们**有时被称作不可见光或黑光。**紫外线**的波长在 12 毫微米—390 毫微米之间。波长在 300 毫微米—390 毫微米之间的**紫外线**会把人的皮肤晒黑。**它们**还能把皮肤中的某种化学成分变成维生素 D。另外，185 毫微米—300 毫微米之间的**紫外线**还可用来杀灭有害细菌。

分析：这个段落的主题是 ultraviolet rays，翻译时突出了这一主题。因此译文的每一句话始终贯穿主题，使它成为连接句与句的纽带。

▶ 2.1.2 从主题入手——突出主次信息

英语作者在表达一个重要信息时，往往会把核心信息放在突出位置。因此，抓住主题句等于掌握了整个段落的纲。在此基础上进行翻译，可以保证段内主要内容不偏离原文，译文主次分明。

There is a difference between science and technology. Science is a method of answering theoretical questions; technology is a method of solving practical problems. Science has to do with discovering the facts and relationships between observable phenomena in nature and with establishing theories that serve to organize these facts and relationships; technology has to do with tools, techniques, and procedures for implementing the findings of science. Another distinction between science and technology has to do with the progress in each.

译文：科学与技术之间有区别。**首先**，科学是回答理论问题的方法，而技术是解决实际问题的方法；**其次**，科学与发现自然界可观察现象之间的事实和关系以及构建它们的理论框架有关，而技术与工具、方法和完成科学发现的过程有关；**最后**，它们各自的进展不同。

分析：本段第一句为主题句，说明"科学与技术之间有区别"，然后通过反复以科学和技术为各句的基本主题展开说明了区别之所在。因此，翻译时在展开部分增加逻辑连接词（首先、其次、最后），使译文结构清晰，主次分明。

当然，英语中也有"主题隐含式"段落，即无主题句的段落，但文中所有的事实和细节形成一个整体，整个段落由一个隐含的思想所统一，其中心思想可以通过段落中的每个句子体现出来。

As traffic congestion spreads, increasing amounts of time and fuel are wasted. More fumes are released into the air, increasing the likelihood that cities will be covered by smog. The lives of sufferers from chest diseases are endangered. Others find that their eyes water and they have a tickling sensation in the nose. The smell of exhaust fumes increasingly covers up nicer smells and scents that are part of the pleasure of life.

译文： 随着交通阻塞现象的蔓延，越来越多的时间和燃料被浪费。更多的废气被排放到空气中，城市被烟雾弥漫的可能性增大。患胸部疾病的病人的生命受到危害。其他人则出现眼睛流泪、鼻子发痒的症状。生活中各种本可以给人带来愉悦感的香气和味道逐渐被废气的气味所覆盖。

分析： 从时间、燃料的浪费，到破坏空气的清洁、危及病人生命等，作者一一列举了城市交通阻塞带来的影响。本段的主题句可归纳为：Traffic congestion is a big problem.，译文主次信息清晰可辨。

2.2 注意段落句间的衔接与连贯

我们知道，词汇和词组构成句子，句子与句子构成段落，所以构成段落的句子之间必定存在着某种逻辑上的联系。而衔接和连贯就是语句成段的重要保证。只有语义连贯并有所衔接的语句才能表达符合逻辑的思想，传达内涵完整的信息。

▶ 2.2.1 衔接的翻译方法

衔接是段落的重要特征。段落内句子之间需通过一定的衔接手段，将句子与句子有逻辑地组织起来，构成一个完整或相对完整的语义单位。段落衔接主要采用语法和词汇手段。试分析例2的段落。

Ultraviolet rays are waves similar to light which are just beyond the violet end of the visible light spectrum. *Ultraviolet rays* are sometimes called invisible light or black light because we cannot see *them* with the human eye. *Ultraviolet rays* have wavelengths between 12 and 390 millimicrons. *The rays* with wavelengths between 300 and 390 millimicrons make people's skin tan. *They* also change a chemical in the skin into Vitamin D. *Those* between 185 and 300 millimicrons are used for killing harmful bacteria.

分析：本段使用了语法手段中的照应关系（如 them，the rays，they，those 等），也使用了词汇手段中的复现关系（如反复使用关键词 ultraviolet rays 等）。衔接手段的使用使本段文字内容紧凑，条理清楚，连贯性强。

英语是重视衔接的一种语言，语句之间需要词汇加以过渡和引导；而汉语在很多情况下没有词汇衔接语句，是通过语句之间的逻辑关系组合成段的。因此在翻译过程中，我们一定要兼顾两种语言的不同特点，通过衔接手段，把握原文的逻辑思路，对原文的衔接方式进行必要的转换和变通。具体翻译方法如下：

（1）突出照应词汇

照应是表示语义关系的一种语法手段，翻译时突出照应使段落更显其结构上的衔接和语义上的连贯。

Rontgen went on to try to find out more about *these strange invisible rays*. It appeared that *they* travel in a straight line, as light rays travel; but that *they* are not bent, as light is, when passing through glass or water. He gradually came to realize that *X-rays*, like light waves, are electromagnetic radiations but that *their* wavelengths are extremely short. This is why *they* are invisible to our eyes.

译文：伦琴对 X 射线作了进一步研究。他发现，X 射线和光线一样都是直的，但当它们穿过玻璃或水时，不会像光线那样发生折射。他逐渐意识到，X 射线和光波一样，都属于电磁辐射，只是 X 射线的波长特别短，这就是肉眼看不见它们的原因。

分析：段落中出现了不少照应词汇，如 these strange invisible rays，they

Unit Nine Astronautics

和 their 等，通过这些词，我们能连贯地把握段落的逻辑架构。翻译时在每句中突出照应词，使段落的主旨更加清楚，句际关系更加流畅。

（2）重复关键词

段落的语义连贯可以通过词汇衔接手段予以实现，如复现关系（reiteration）。复现关系主要是通过反复使用关键词、同义词、近义词、上义词、下义词、概括词等手段体现的。

一般情况下，我们尽量避免重复使用同一个词，因为这样会使文章显得单调。但是科技文章重复现象很多，因为这种文章需要的是科学性、准确性，而不是灵活多变的选词。

Clouds can greatly affect the temperature of the earth's surface. When there are many clouds in the sky, all of the sun's rays cannot reach the earth. The cloudy day, then, will be cooler than the cloudless day. Clouds also prevent the earth from cooling off rapidly at night. For this reason, countries such as the British Isles, which are often covered by clouds, have a relatively constant temperature. The weather in these cloudy areas is neither very hot in summer nor very cold in winter. On the other hand, places such as deserts, which have few or no clouds, have very sharp variations in temperature—between night and day as well as between summer and winter.

译文：云能大大影响地表温度。白天多云时，太阳的射线不能到达地面，因而比少云时凉快。夜间云能阻止地面散热过快。由于这个原因，常处于多云状态的不列颠群岛上的国家，温度相对恒定。这些地区冬暖夏凉。然而，少云或无云的沙漠地区昼夜和冬夏温差变化极大。

分析：本段主题谈论"云"对地表温度的影响。因此，翻译时，关键词"云""多云""少云""无云"要始终贯穿段落的每一句话，使语句上下衔接且连贯。

（3）利用词汇链（lexical chain）

段落的语义连贯还可以通过同现关系来实现。同现关系（collocation）

指的是词语在段落中同时出现的倾向性或可能性。如例7，围绕 input device 一词，人们常联想到 magnetic tape、disk、terminal、keyboard、tape drive、disk drive、medium、printer、CRT display screen 等词，也就是说，这些词可能会在段落中与 input device 一起出现。它们属同一个词汇链，因此能在段落中起到衔接上下文的作用。翻译时，利用词汇链关系，可以在词汇上排除歧义，保证译文内容不偏离原文。

  Some of the most common methods of *inputting information* are to use *magnetic tape*, *disks*, and *terminals*. The computer's *input device* (which might be a *keyboard*, a *tape drive* or *disk drive*, depending on the *medium* used in inputting information) reads the information into the computer. For *outputting information*, two common devices used are a *printer* which prints the new information on paper, and a *CRT display screen* which shows the results on a TV-like screen.

  输入信息的一些最普通的方法是使用**磁带**、**磁盘**和**终端**。计算机的**输入装置**（依据输入信息时使用的**媒体**，可能是**键盘**、**磁带机**或**磁盘驱动器**）把信息读入计算机内。对于**输出信息**，有两种常用的装置：把新信息打印在纸上的**打印机**，以及在类似电视的荧屏上显示结果的**阴极射线管显示屏**。

（4）增添连接词

增添衔接性词语是保证语义贯通的一个常用方法。

  For the moment, drilling has stopped. One of the propulsion motors that controls the automatic thrusters has malfunctioned.

译文：这时，钻探停下了。**原来**控制自动推进器的马达中有一台出了故障。

分析：从段落概念来说，这两句话构成一个语义整体，后一句对前一句作了补充说明。翻译时，添加"原来"以显示段落内含有的因果关系。

（5）拆句分译

由于英语和汉语在句型结构上的不同，英译汉时，可以打破原文中与汉

Unit Nine Astronautics

语不相适应的结构，拆句分译，将英语逻辑严谨的复合结构按汉语的行文方式拆开来译，按汉语习惯语序铺排。

Computer languages may **range from** detailed low level close to that immediately understood by the particular computer, **to** the sophisticated high level which can be rendered automatically acceptable to a wide range of computers.

译文：计算机语言有低级的，也有高级的。前者比较详细，很接近于特定计算机直接能懂的语言；后者比较复杂，适用范围广，能自动为多种计算机所接受。

分析：如果不采用分译，把 range from...to... 译成"范围从……到……"，则会显得行文累赘、脉络不清。译文打破了原文的句子结构，使用了"前者……后者……"这种结构。这样处理，逻辑合理，且使语言清晰而流畅。

Once combustion starts, it should be carried through the mixture very rapidly, and this is assisted by making the clearance space above the piston as small as possible, and by careful design of the cylinder head. Rapid propagation of the flame through the compressed gas is also assisted by creating turbulence in the gas.

译文：燃烧一旦开始，就应当迅速地使全部混合气燃烧起来。要做到这一点，应当使活塞上方的间隙尽可能小，并要精心地设计汽缸盖；使压缩气体产生涡流，也有助于火焰在其中迅速蔓延。

分析：译文采用拆句分译的方法，将原来的句子按照汉语习惯重新安排，条理清晰，逻辑严密。

（6）调整语序

根据段落的需要，翻译有时会调整语序，使行文更为紧凑和连贯。

Before long, plastic may transform its image from eco-villain to environment hero, thanks to "smart" plastics now in development. These could allow vehicles to eliminate ozone-depleting emissions and help windows store and make use of the sun's heat. Perhaps by then, the name itself will morph from a pejorative jab into a genuine compliment.

译文：目前,"智能"塑料正在开发之中,这种材料不仅能消除汽车排放的破坏臭氧的废气,还可以让窗户储存和利用太阳的热量。不久的将来,塑料的形象将从生态魔鬼摇身一变为环保英雄。也许到那时,塑料一词的含义将从恶意的挖苦转成真正的赞美。

分析：本段落呈现了一种因果关系：因为"智能"塑料的开发,所以塑料的形象将从反面转向正面。因此,按照汉语先因后果的思维习惯,在翻译时可以对语序进行适当的调整,使这种因果关系更加清晰。

Extracting pure water from the salt solution can be done in a number of ways. One is done by distillation, which involves heating the solution until the water evaporates, and then condensing the vapor. Extracting can also be done by partially freezing the salt solution. When this is done, the water freezes first, leaving the salt in the remaining unfrozen solution.

译文：从盐水中提取纯水的方法有很多种。一种是加热蒸馏法,另一种是局部冷冻法。加热蒸馏法是将盐水加热,使水分蒸发,然后再使蒸汽冷凝成水。局部冷冻法是使盐水部分冷却,这时先行冷冻的是水,盐则留在未冻结的液体中。

分析：为了段落的衔接,翻译时可以不拘泥于表面词义的先后出现顺序,将后句的内容提到前句交代。

▶ 2.2.2 连贯的翻译方法

连贯是指各句之间必须衔接自然。段落整体结构连贯,不仅需要句子做到衔接自然、有条不紊,而且句子之间的关系还要合乎逻辑。从翻译的角度来考虑,段落翻译中连贯性的问题是如何保证段落中语义逻辑的一致。具体翻译方法如下：

（1）使用逻辑连接词

所谓逻辑连接词是体现逻辑思维的句子与句子之间的连接手段,包括表示时空关系、列举、引申、转折、推论等的连接词。为了使译文语义连贯,

Unit Nine Astronautics

我们应该弄清楚句间的逻辑意义，再根据汉语的行文习惯保留或略去连接词。

With few exceptions, ***however***, this is not the case with independent voltage and current sources. ***Although*** an actual battery can often be thought of as an ideal voltage source, other non-ideal independent sources are approximated by a combination of circuit elements. Among these elements are the dependent sources, which are not discrete components as are many resistors and batteries ***but*** are in a sense part of electronic devices like transistors and operational amplifiers. ***But*** don't try to peel open a transistor's metal so that you can see a little diamond-shaped object. The dependent source is a theoretical element that is used to help describe or model the behavior of various electrical devices.

译文：然而，除了极少数几个例子外，对于独立的电压源和电流源来说，情况就并不是这样。**虽然**电池本身往往可以看成是理想电压源，**但是**也可以用一组电路元件做成类似非理想的独立电源。这些元件中有非独立源，它们并不像许多电阻器和电池那样属于分立部件，**而**在某种意义上属于像晶体管和运算放大器这样的电子器件的一部分。**不过**不要试着打开晶体管壳，因为所有你能看见的不过是一个小小的菱状物。非独立源是用来帮助描述各种电器件的性能或对其进行建模的一种理论元件。

分析：本段落句间的语义连贯主要通过使用表示转折的连接词（如 however、although、but 等）达到的，所以句子衔接自然，整段文章连贯协调。

（2）重视语段内的逻辑关系

段落是按一定逻辑组织构成的，翻译时，可根据译文的行文习惯，调整语段内的逻辑关系，删去重复赘述，重新组织译文，使含糊不清的概念明晰起来。这样可以使译文言简意赅、表达确切，同时又符合汉语表达习惯。

In an age of super sonic airliners it is difficult to realize that ***at the beginning of the twentieth century*** no one had ever flown in an aeroplane. However, people were flying in balloons and airships. The airship was based on the principle of the semi-rigid structure. ***In 1900*** Ferdinand von

Zeppelin fitted a petrol engine to a rigid balloon. This craft was the first really successful steerable airship. *In 1919* an airship first carried passengers across the Atlantic, and *in 1929* one traveled round the world. During this time the design of airships was constantly being improved and *up to 1937* they carried thousands of passengers on regular transatlantic services for millions of miles.

译文：在超音速飞机时代，很难意识到 **20 世纪初期**的人们还没法驾驶飞机飞行。然而，那时人们驾驶气球和飞艇飞行。飞艇是根据半刚性结构原理飞行的。**1900 年**，斐迪南·冯·齐柏林将汽油引擎安装在硬质气球上。这是第一艘真正成功的可驾驶的飞艇。**1919 年**，飞艇首次运送乘客飞越大西洋，并**于 1929 年**环游世界。在此期间，飞艇的设计不断得到改进，**到 1937 年**，飞艇可以运送乘客数千人，飞行数百万英里，开辟了定期越洋航班服务。

分析：在这一段落里，作者用时间顺序来描述飞艇的发展史。整个段落使用了很多时间状语，如 at the beginning of the twentieth century、in 1900、in 1919、in 1929、up to 1937 等，从而形成非常清晰的飞艇发展的时间框架。翻译时，时间状语将各句衔接起来，译文段落的组织环环相扣、紧密和谐，反映了原文的脉络。

The Moto 1100 is a small family car. It has a small engine which is *in the front*. The engine has *a capacity of 1,100 cubic centimeters*. It is a *front wheel drive* car. The gear lever is *on the floor*. There are seats for four or five people. It has four front forward gears and a reverse. It has *a maximum speed of about 130km an hour*. One advantage of this car is that it has *a very low fuel consumption*.

译文：1100 型汽车是一种小型家用小汽车。车的**前部**装有一个小型发动机，其**排量为 1,100 立方厘米**。该车为**前轮驱动**，变速杆装在车内**地板上**，且有四五个座位。这种车还有四个前进挡和一个倒车挡。**最快时速约为 130 公里**，其优点之一是耗油很少。

分析：本段描写 1100 型小汽车的特征及空间构造。翻译时，从段落整体考虑，重新组织和调整，使译文更加准确，更富有条理性和逻辑性。

Radial bearings which carry a load acting at right angles to the shaft

Unit Nine Astronautics

axis, and thrust bearings which take a load acting parallel to the directions of shaft axis—are two main bearings used in modern machines.

译文：承受的荷载与轴心线成直角的是径向轴承，而承受的荷载与轴心线相平行的是止推轴承，这是现代机器上使用的两种主要轴承。

分析：文中两个 which 在译文中都没有体现，译文把两个带定语从句的主语，分别译成两个短句，再借助代词"这"来引出句子的主干，这种相反的叙述层次，使译文一气呵成，连贯紧密。

Quiz

1. **Determine the topic of the following paragraph, and then translate the whole paragraph into Chinese.**

   A transmitter commonly consists of several parts. It is an equipment to send out radio waves. The use of a telecommunications transmitter is to transmit intelligence by radio. To transmit intelligence by radio, it is necessary to generate high-frequency signals, because radio waves can be sent out only if the frequency is high.

2. **Translate the following paragraphs into Chinese.**

   (1) There are two factors which determine an individual's intelligence. The first is the sort of brain he is born with. Human brains differ considerably, some being more capable than others. But no matter how good a brain he has to begin with, an individual will have a low order of intelligence unless he has opportunities to learn. So the second factor is what happens to the individual—the sort of environment in which he is reared. If an individual is handicapped environmentally, it is likely his brain will fail to develop and he will never attain the level of intelligence of which he is capable.

   (2) When Rontgen wrote an account of what he had discovered, he called these new rays X-rays, for X is a symbol often used

for something which is not yet understood. Other scientists called them Rontgen rays in honor of the man who first found them, but X-ray is the name always used now.

(3) After the bricks are pressed they may be cooled for storage or taken directly to ovens for tempering. Tempering of the pitch-bonded brick has been found to improve several brick characteristics.

(4) Maglev vehicles comprise a minimum of two sections, each with approximate 90 seats on average. According to application and traffic volume, trains may be composed of up to ten sections (two end and eight middle sections).

(5) So electronics is the basis of all telecommunications systems. The radio valves which contained a gas at very low pressure are now being replaced by solid-state semiconductor devices called transistors which are much smaller and have a longer life than gas valves. Since the last war a whole new industry concerned with the manufacture and exploitation of solid-state electronics has been established. One new branch of this field has led to the development of computers which contain enormous numbers of transistors, minute magnetic cores and other new electronic solid-state devices.

Part B Reading

The Scientific Exploration of Space

Since the War, and particularly during the last few years, a rapidly growing amount of effort has been devoted to the use of high-power rockets to carry instruments up to great heights above the Earth, to launch artificial satellites and deep space **probes**.[1] We have pointed out how much has been and can still be done from the earth's surface. Why then all these concentrations on the use of rockets?

One of the main reasons is that our atmosphere, while beneficial for life in general, prevents us from seeing the universe in any but a very restricted range of light—almost entirely **confined to** visible light and to a relatively restricted range of radio waves, in fact.[2] We must make observations from outside the atmosphere to study the ultraviolet light, X-rays, infrared rays and all those radio waves that cannot **penetrate** through our atmosphere. With instruments in artificial satellites circulating at heights of over 200 miles such observations can be made. What they will record, we do not know—if we did, it would not be worth going to all this trouble—but there is **scope** here for astronomical studies for generations to come.[3]

This is only one of the many major new possibilities for scientific research which are **opened up** by the development of rocket vehicles in the study of the earth's outer atmosphere, in **meteorology**, in the study of the space between the earth and the planets, and so on. There are four main categories of vehicles involved in this work, which has been called space research. First, there are vertical sounding rockets, which can go up to 1,000 miles or more, but are of most use below 200 miles. These rockets simply rise to the top of their **trajectory** and fall back to earth. Next, we have the artificial satellites revolving round the earth in elliptical paths, never penetrating closer than about 150 miles or so. A speed as high as

18,000 miles per hour must be given to a body to launch it as a satellite. If the satellite orbit is very **elongated**, so that it passes out to distances several times the earth's radius (4,000 miles), we have a deep space probe.[4] The greater the launching speed, the greater the penetration into space before they return to earth. Eventually, when the speed reaches 25,000 miles per hour, the probe never returns but becomes a satellite of the sun, an artificial planet. Probes may be specially directed to pass near the moon, or hit the moon or become satellites of the moon. These are the lunar probes, of which there have been a number of examples recently.[5]

From the scientist's point of view, all these vehicles play a valuable part. The availability of artificial satellites does not make vertical sounding rockets **obsolete** and this is even more true of deep space and lunar probes in relation to artificial satellites. To the scientist the value of any particular launching is the success of the experiment concluded, not just the distance reached from the earth. Nor is he concerned with putting men in the vehicle, for the instruments can be made to operate automatically and to send back their readings to earth—even over distances of millions of miles—as coded radio signals.[6]

New Words and Expressions

probe 探测器
confine to 限制
penetrate 穿过
scope（发挥能力的）余地；机会
open up 开辟

meteorology 气象学
trajectory 轨道
elongate 拉长；(使) 延长
obsolete 过时的

Notes

1. 句中 the War 指的是 Second World War，名词 the use 转译成动词 to use，与 to launch 成并列结构。

2. 句中 that 引导表语从句，在全句中做表语。句中 prevents us from seeing the universe in any but a very restricted range of light 的意思是 we can see the universe only in a very restricted range of light，译为：……我们只能在一个非常有限的光域内看见宇宙……。

3. 句中 they 指代的是上一句中的 instruments，翻译时应予以表述。句中 scope 意为"（发挥能力的）余地；机会"。全句译为：我们不知道这些仪器会记录些什么，如果知道的话，就不值得找这么多麻烦了。但是这个天文学研究领域有待于以后几代人进行研究。

4. 句中 so that 引导目的状语从句。全句译为：如果卫星轨道特别长，超过地球半径（4,000 英里）数倍时，我们就采用外层空间探测器。

5. 本句采用合句法和断句法翻译。将句子分成两部分，前一部分 These are the lunar probes 与上一句合译，后部分 of which there have been a number of examples recently 单独成句。这两句译为：探测器可以经过专门导向从月球附近掠过，也可撞击月球或成为月球的卫星，这就是月球探测器。近来我们已经有了许多这类探测器的实例。

6. 句中 nor 是否定代词，在句首时，句子须倒装。

Exercises

1. Find out the English equivalents of the following Chinese terms from the passage.

   （1）外层空间探测器　　　（2）月球探测器
   （3）火箭运载工具　　　　（4）探测火箭
   （5）椭圆形轨道　　　　　（6）人造卫星
   （7）紫外线　　　　　　　（8）红外线
   （9）X 射线　　　　　　　（10）无线电波

2. Translate the following sentences into Chinese.

   (1) Since the War, and particularly during the last few years, a rapidly growing amount of effort has been devoted to the use of high-power rockets to carry instruments up to great heights above the Earth, to

launch artificial satellites and deep space probes.

(2) One of the main reasons is that our atmosphere, while beneficial for life in general, prevents us from seeing the universe in any but a very restricted range of light—almost entirely confined to visible light and to a relatively restricted range of radio waves, in fact.

(3) What they will record, we do not know—if we did, it would not be worth going to all this trouble—but there is scope here for astronomical studies for generations to come.

(4) This is only one of the many major new possibilities for scientific research which are opened up by the development of rocket vehicles in the study of the earth's outer atmosphere, in meteorology, in the study of the space between the earth and the planets, and so on.

(5) If the satellite orbit is very elongated, so that it passes out to distances several times the earth's radius (4,000 miles), we have a deep space probe.

(6) The availability of artificial satellites does not make vertical sounding rockets obsolete and this is even more true of deep space and lunar probes in relation to artificial satellites.

(7) Nor is he concerned with putting men in the vehicle, for the instruments can be made to operate automatically and to send back their readings to earth—even over distances of millions of miles—as coded radio signals.

Part C Extended Reading

Introduction to Space Exploration

Space exploration is the **quest** to use space travel to discover the nature of the universe beyond Earth. Since ancient times, people have dreamed of leaving their home planet and exploring other worlds. In the later half of the 20th century, that dream became reality. The space age began with the launch of the first artificial satellite in 1957. A human first went into space in 1961. Since then, astronauts and **cosmonauts** have **ventured** into space for ever greater lengths of time, even living aboard orbiting space stations for months on end. Two dozen people have circled the moon or walked on its surface. At the same time, **robotic** explorers have journeyed where humans could not go, visiting all but one of the solar system's major worlds. Unpiloted spacecraft have also visited a host of minor bodies such as moons, **comets**, and asteroids. These explorations have sparked the advance of new technologies, from rockets to communications equipment to computers. Spacecraft studies have **yielded** a **bounty** of scientific discoveries about the solar system, the Milky Way Galaxy, and the universe. And they have given humanity a new perspective on the earth and its neighbors in space.

The first challenge of space exploration was developing rockets powerful enough and reliable enough to boost a satellite into orbit. These **boosters** needed more than **brute force**, however; they also needed guidance systems to steer them on the proper flight paths to reach their desired orbits. The next challenge was building the satellites themselves. The satellites needed electronic components that were lightweight, yet durable enough to **withstand** the acceleration and vibration of launch. Creating these components required the world's **aerospace** engineering facilities to adopt new standards of reliability in manufacturing and testing.

On earth, engineers also had to build **tracking stations** to maintain radio communications with these artificial "moons" as they circled the planet.

Beginning in the early 1960s, humans launched probes to explore other planets. The distances traveled by these robotic space travelers required travel times measured in months or years. These spacecraft had to be especially reliable to continue functioning for a decade or more. They also had to withstand such hazards as the radiation belts surrounding **Jupiter**, particles orbiting in the rings of **Saturn**, and greater extremes in temperature are faced by than spacecraft in the **vicinity** of Earth. Despite their great scientific returns, these missions often came with high **price tags**. Today the world's space agencies, such as the United States National Aeronautics and Space Administration (**NASA**) and the European Space Agency (ESA), strive to conduct robotic missions more cheaply and efficiently.

It was inevitable that humans would follow their unpiloted creations into space. Piloted space flight introduced a whole new set of difficulties, many of them concerned with keeping people alive in the hostile environment of space. In addition to the vacuum of space, which requires any piloted spacecraft to carry its own atmosphere, there are other deadly hazards: solar and **cosmic** radiation, **micrometeorites** (small bits of rock and dust) that might **puncture** a spacecraft **hull** or an astronaut's pressure suit, and extremes of temperature ranging from **frigid** darkness to **broiling** sunlight. It was not enough simply to keep people alive in space—astronauts needed to have a means of accomplishing useful work while they were there. It was necessary to develop tools and techniques for space navigation, and for conducting scientific observations and experiments. Astronauts would have to be protected when they ventured outside the safety of their pressurized spacecraft to work in the vacuum. Missions and hardware would have to be carefully designed to help insure the safety of space **crews** in any foreseeable emergency, from liftoff to landing.

The challenges of conducting piloted space flights were great enough for missions that orbited Earth. They became even more **daunting** for the Apollo missions, which sent astronauts to the moon. The achievement of sending astronauts to the lunar surface and back represents a summit of human space flight.

After the Apollo program[1], the emphasis in piloted missions shifted to long-duration space flight, as pioneered aboard Soviet and US space stations. The development of reusable spacecraft became another goal, keeping rise to the US space shuttle fleet. Today efforts focus on keeping people healthy during space missions lasting a year or more—the duration needed to reach nearby planets—and on lowering the cost of sending satellites into orbit.

New Words and Expressions

quest 探求，搜寻
cosmonaut（尤指苏联的）宇航员
venture 冒险
robotic 机器人的
comet 彗星
yield 产生；带来
bounty 慷慨的赠予，恩惠；赠物，赠礼
booster（加速）推进器
brute force 蛮力
withstand 顶得住，经受住
aerospace 航空和宇宙航行空间（指地球大气层及其外面的宇宙空间）

tracking station 跟踪站
Jupiter 木星
Saturn 土星
vicinity 附近地区，近邻
price tag 价目标签
NASA 美国航空航天管理局
cosmic 宇宙的
micrometeorite 微陨石，陨石微粒
puncture 刺穿；戳破
hull（飞船的）船身
frigid 寒冷的
broiling 火辣辣的
crew 全体人员
daunting 令人畏惧的

Notes

1. Apollo program，"阿波罗"登月计划。1969年7月20日，美国航天员阿姆斯特朗和奥尔德林驾驶"阿波罗11号"飞船的登舱降落在月赤道附近的静海区，首次实现了人类登上月球的梦想。这是一次震动全球的壮举，也是世界航天史上具有重大历史意义的成就。

Exercises

1. Translate the following sentences into Chinese.

(1) Since ancient times, people have dreamed of leaving their home planet and exploring other worlds. In the later half of the 20th century, that dream became reality. The space age began with the launch of the first artificial satellite in 1957. A human first went into space in 1961. Since then, astronauts and cosmonauts have ventured into space for ever greater lengths of time, even living aboard orbiting space stations for months on end.

(2) These explorations have sparked the advance of new technologies, from rockets to communications equipment to computers. Spacecraft studies have yielded a bounty of scientific discoveries about the solar system, the Milky Way Galaxy, and the universe. And they have given humanity a new perspective on the earth and its neighbors in space.

(3) In addition to the vacuum of space, which requires any piloted spacecraft to carry its own atmosphere, there are other deadly hazards: solar and cosmic radiation, micrometeorites (small bits of rock and dust) that might puncture a spacecraft hull or an astronaut's pressure suit, and extremes of temperature ranging from frigid darkness to broiling sunlight.

(4) Today efforts focus on keeping people healthy during space missions lasting a year or more—the duration needed to reach nearby planets—and on lowering the cost of sending satellites into orbit.

Unit Ten
Civil Aviation

Part A Lecture

科技文章的翻译

科技文章包括科普文章、科技论文、科技报道、实验指示与实验报告、科技发展的历史等。科技文章种类不同，每个种类都有各自的特点。有些科技文章（如科技论文、各种学术文献）描述高度复杂的活动或关系，所以结构严密，句子较长而且复杂，逻辑性很强。有些科技文章（如实验指导书等文字材料）内容一般为解释、规定、建议等，所以结构简单，句子短小精悍。尽管如此，科技文章仍具有一些共同点，即条理分明，层次清楚，语言简练，合乎逻辑。

因此，进行科技文章翻译时，首先要注意文章的逻辑关系，明确翻译的对象属于哪种文章，才能准确把握该文体结构的特征和文章的语言逻辑。其次要注意文章的语篇衔接连贯。语篇衔接通过词汇或语法手段使文脉相通，形成语篇的有形网络。语篇连贯以信息发出者和接受者双方共同了解的情景为基础，通过推理来达到语义的连贯，这是语篇的无形网络。充分利用语篇的叙事次序（时间顺序、空间顺序等）和逻辑连接词有助于我们对文章的整体把握，做到传意达旨。下面分别分析科普文章、科技论文、科技报道、实验指示与实验报告的文体特点及其翻译方法。

1 科普文章的文体特点及其翻译

科普文章是普及科学知识的读物，它包罗万象，涉及天文、地理、物理、化学、生物、医学等方面知识，其形式也多样化，有科学小丛书、百科全书、科普文摘、科学家传记等。

科普文章内容上具有常识性、知识性和趣味性的特点；语言上通俗易懂，深入浅出，语句简短，多用普通词汇。在翻译这类文章时应忠实于原文风格，用生动灵活、浅显易懂的汉语普及科学知识。

Unit Ten　Civil Aviation

　　Food quickly spoils and decomposes if *it* is not stored correctly. Heat and moisture encourage the multiplication of microorganisms, and sunlight can destroy the vitamins in such foods as milk. Therefore, most foods should be stored in a cool, dark, dry place which is also clean and well ventilated.

　　Foods that decompose quickly, such as meat, eggs, and milk should be stored in a temperature of 5℃–10℃. In *this* temperature range, the activity of microorganisms is considerably reduced. In warm climates, this temperature can be maintained only in a refrigerator or in the underground basement of a house. In Britain, for six months of the year at least, *this* temperature range will be maintained in an unheated room that faces the north or the east. Such a room will be ideal for food storage during the winter months provided that *it* is well ventilated.

分析：从文章的逻辑关系来看，这篇文章组织严密，逻辑性强。文章的层次清楚，第一段第一句话概括了该段的中心思想，第二段具体说明了保存食物的要求。

　　从文章的语篇衔接连贯角度来看，在衔接方面，使用了语法手段的"照应"，第一段代词 it 指代 food，第二段 it 指代 room，因为它们在上下文的意思非常清楚，故略去不译。而第二段中指代温度范围的 this，因为译文中需要强调，故保留其意。

　　文章中还使用了词汇手段"词语的复现关系"，如 food、decompose、store（storage）都出现两次以上；以意义相同或相近的词组的形式出现，如文章前面用 spoil、decompose，后面扩展为 the multiplication of microorganisms、the activity of microorganisms、destroy the vitamin 等；前面用 cool，后面用各种方式表达同一概念：in a temperature of 5℃–10℃、in a refrigerator、in the underground basement of a house、in an unheated room that faces the north or the east、during the winter months。通过词汇在意义上的衔接把全篇文章的各部分紧紧地联系在一起，使文章结构紧凑，前后呼应。

译文：食物如果保存不当很快会腐烂。温度和潮气都会助长微生物的繁殖，阳光会破坏牛奶一类食品的维生素成分。所以，大多数食物应该保存在既凉爽、黑暗、干燥又干净且通风良好的地方。

　　腐烂变质很快的食品，如肉类、蛋类和奶类，应该在 5℃ 至 10℃ 范围保存。因为在此温度范围内，微生物的活动大大减少。在气候温暖的情况下，只有在冰箱或房屋的地下室才能保持这个温度。在英国，一年至少有 6 个月时间，在朝北或朝东没有暖气的房间里，这个温度范围可以保持。在通风良好的条件下，这样的房间适合冬季保存食物。

Natural Energy

Energy in nature comes in many different forms. Heat is a form of energy. A lot of heat energy comes from the sun. Heat can also come from a forest fire or, in much smaller quantities, from the warm body of a mouse. Light is another form of energy. It also comes from the sun and from the stars. Some animals and even plants produce small amounts of light energy. Radio waves and ultraviolet rays are other forms of energy. Then there is electricity, which is yet another sort of energy.

All these different forms of energy can be changed, one into another. Thinking of lightning. All the electrical energy in it is gone in a flash—changed into brilliant light which you can see, into heat which burns whatever is struck by the lightning, and into sound which you can hear as thunder.

Much of the energy we use at home comes from electricity. Most of the Earth's energy—wind, waves, heat and light—comes from the sun. The sun itself is powered by nuclear energy.

There are some things about energy that are difficult to understand. The fact that it constantly changes from one form to another makes energy rather like a disguised artist. When you think you know what energy is, suddenly it has changed into a totally different form. But one thing is certain: energy never disappears and, equally, it never appears from nowhere. People used to think that energy and matter are two different things, but now we know they are interchangeable. Tiny amounts of matter convert into unbelievably huge amounts of nuclear energy. The sun produces nuclear energy from hydrogen gas and, day by day, its mass gets less, as matter is converted to energy.

分析： 这篇文章语言直白、结构简单、句子短小。文章给出了大量的实例，多次使用了比喻等修辞方法，使文章通俗易懂；道理深入浅出，易于接受。

译文： 自然界的能量

自然界的能量有许多不同的形式。热能就是一种形式的能量。热能有许多是来自太阳的。森林大火也可以产生热能，甚至一只老鼠温暖的身体也可以产生少许的热能。光是能量的另一种形式，也是来自太阳和星星。一些动物甚至植物也可以产生少量的光能。无线电波和紫外线也是能量的形式。另外还有电能也是一种能量的形式。

所有这些不同形式的能量都可以相互转换。就拿闪电来说，里面所有的电能都在一道闪光中放掉了——转变为可以看见的耀眼光芒，转变为可以烧毁所有被闪电击中的物体的热能，也转变为可以听得见的雷声。

我们在家里使用的很多能量来自电。地球上的大多数能量——风、浪、热、光——来自太阳。而太阳本身的能量是由核能产生的。

有关能量的一些事情很难理解。能量不断地从一种形式转变为另一种形式，就像一位伪装艺术家一样。当你自认为了解它的时候，它突然又变成了另一种完全不同的形式。但是有一点是肯定的：能量永远不会消失，同样，它也不会无端地产生。过去，人们认为能量和物质是两种完全不同的东西，现在我们知道，能量和物质是可以相互转换的。微量的物质可以转换为令人难以置信的巨大核能。太阳利用氢气制造核能，随着物质转化为能量，其质量日复一日在减小。

❷ 科技论文的文体特点及其翻译

科技论文是在科学实验和科学研究的基础上对科学技术领域内的某些现象与问题进行科学的分析和阐述，从而揭示这些现象和问题的本质及其规律的文章。按写作目的，科技论文可分为期刊论文、学位论文和会议论文。按论文内容，科技论文可分为实验型、理论型和综述型三种。

科技论文是科技研究人员研究成果的直接记录，因此论文内容专业性强，文字规范、严谨。论文侧重叙事和推理，具有很强的逻辑性：结构严谨、层次分明、前提完备、概念确切、推理严密、分析透彻、判断准确。

科技论文有两大文体特点。其一是文体正式：科技论文采用书面文体，用词准确、语气正式、语言规范，避免口语化用词，不用或少用第一、第二人称代词，行文严谨简练。其二是高度的专业性：科技论文均有一个专业范围，因此，专业术语是构成科技论文的语言基础，其语义具有严谨性和单一性等特点。

英语科技论文虽然在目的性上与用其他语言形式撰写的科技论文是相同的，但是由于文化的差异和英语语言本身的特点，英语科技论文还有自己的语言特征、文体要求和格式变化。

因学科不同，且研究项目、过程和结果不同，科技论文可以有多种体例结构。常见的科技论文结构非常规范，一般包括以下内容：

（1）论文题目　　　　Title
（2）作者　　　　　　Author
（3）摘要　　　　　　Abstract
（4）关键词　　　　　Key words
（5）正文　　　　　　Body
（6）致谢　　　　　　Acknowledgments（可空缺）
（7）参考文献　　　　References
（8）附录　　　　　　Appendix（可空缺）
（9）作者简介　　　　Resume（视刊物而定）

其中正文为论文的主体部分。正文又分为若干章节，一篇完整的科技论文的正文部分由以下内容构成：

（1）引言/概述　　　Introduction
（2）文献综述　　　　Literature Review
（3）材料和实验　　　Materials and Experiments
（4）实验结果　　　　Results
（5）讨论　　　　　　Discussion
（6）结论/总结　　　Conclusions

由于目前我们接触的更多是科技论文的摘要，所以本文只涉及摘要的翻译。

摘要是全篇论文的缩影，常常被专业期刊文献杂志编入索引资料或文献刊物，以便学术之间的交流。摘要的英文术语原来有两个词汇，摘要（abstract）和概要（summary）。摘要不是正文的组成部分，但隶属于论文；而概要则是正文的一个组成部分。摘要放在正文前，概要一般安排在正文的最后。

科技论文摘要要求围绕着正文的论题，就研究的目的、方法、结果、结论等主要环节进行概括性介绍；概要是正文各部分内容的综合性复述，因此翻译时要力求准确、清晰和简洁。准确是指内容上要忠实于原文。清晰要求使用标准术语，不用第一、二人称，不混用时态，不使用带有感情色彩和意义不确定的词。简洁指用有限的字数将文章的论题、论点、实验方法、实验结果表达出来。

Composing Letters with a Simulated Listening Typewriter

Abstract:

Background	With a listening typewriter, what an author says would be automatically recognized and displayed in front of him or her. However, speech recognition is not yet advanced enough to provide people with a reliable listening typewriter.
Purpose	An aim of our experiments was to determine if an imperfect listening typewriter would be useful for composing letters.
Method	Participants dictated letters, either in isolated words or in consecutive word speech. They did this with simulations of listening typewriters that recognized either a limited vocabulary or an unlimited vocabulary.
Result	Results indicated that some versions, even upon first using them, were at least as good as traditional methods of handwriting and dictating.
Conclusion	Isolated word speech with large vocabularies may provide the basis for a useful listening typewriter.

使用模拟听写打字机写信

摘要：

（背景）	有了听写打字机，作者说的话就会自动被听辨出来并显示在其面前。然而，语言识别的功能还不够先进，目前还无法给人们提供可靠的听写打字机。
（目的）	本试验的目的是要确定一台尚不完善的听写打字机能否用来写信。
（方法）	试验者口授信函时，有时是一字一字口授，有时是连续的话语口授。他们口授时，听写打字机便做模拟工作；听写打字机有的能识别有限量的词语，有的则能识别无限量的词语。

（结果） | 结果表明，即使第一次使用听写打字机，有一些信的效果就和传统耳听、手写的效果一样好。
（结论） | 具有较大词汇量的逐词口授将为可用听写打字机奠定基础。

Abstract: The goal of the investigation is to experimentally demonstrate the feasibility of cogeneration with the nuclear heating reactor, raise its annual availability factor, expand its application field, therefore the economic viability of nuclear heating reactors would significantly be improved. The 5MW nuclear heating reactor (NHR-5) was used as an energy source, and nuclear energy-electricity transformation was achieved with a specially designed low pressure steam generator set in the secondary loop, and a low pressure two-phase flow turbine, meanwhile the circulation cooling water from the condenser was pumped to the heat grid for house heating. The experiments demonstrate that heat-electricity cogeneration using the nuclear heating reactor with integrated design and natural circulation is technically feasible and the cogeneration system shows excellent operability and safety.

摘要：该研究旨在通过实验论证核供热堆进行（热电）联产的可行性，提高其年运行因子，扩大低温堆的应用领域，从而大幅度改善核供热堆的经济性。实验利用5MW核供热堆（NHR-5）为热源，通过其二回路上特殊设计的低压蒸发器和低压两相透平发电机实现核热—电力转换。同时用冷凝器的循环冷却水向热网用户供热，（实现核供热反应堆的热电联产）。结果表明，一体化自然循环式供热堆用于热电联产在技术上是可行的，整套系统表现出极好的运行性能和安全性能。

以上两例讨论了摘要的翻译，下面请试译概要（报告）。

Surveying Natural Resources

1. This section deals with natural resources such as minerals, water and energy. The United Nations family has considerably increased its activities in the field of natural resources and is intensifying its efforts to secure the application of modern technology to resources development. New

Unit Ten Civil Aviation

approaches are also being explored, such as the preparation of projections combining work on related but different groups of resources.

2. The United Nations Secretariat is organizing training programs in the field of natural resources for personnel from developing countries, mostly through regional and inter-regional seminars. These seminars are also useful means of transferring advances in technology to developing countries. They have covered such topics as cartography in relation to development, desalination, energy policy and geochemical techniques in mineral exploration.

3. The Secretary-General has prepared a five-year survey program which is designed to contribute to the development of natural resources by indicating economic and technologically-advanced approaches to the exploration and assessment of these resources. The proposed program consists of nine surveys in the fields of mineral resources, water resources, energy and electricity, as follows:

(a) In the field of mineral resources, a survey of off-shore mineral potential in developing areas, a survey of world iron-ore resources, a survey of important non-ferrous metals and a survey of selected mines in developing countries with a view to increasing ore reserves and production through the application of modern technology.

(b) In the field of water resources, a survey of water needs and water resources in potentially water-short developing countries and a survey of the potential for development in international rivers.

(c) In the field of energy and electricity, a survey of potential geothermal energy resources in developing countries, a survey of oil shale resources and a survey of the needs for small-scale power generation in developing countries.

4. Each of the nine proposed surveys would have two objectives: first, to provide significant new information, ideas and approaches on the natural resources potential of each developing country concerned, and secondly, to gather data that would produce a world-wide perspective of the long-term potential availabilities and needs in the selected areas. They would also be useful in preparing and selecting projects for submission to multilateral or bilateral sources of technical and financial aid.

调查自然资源

1. 本文所述及的，是有关矿物、水和能源等自然资源方面的问题。联合国成员国已大大增加了在自然资源方面的各项活动,并正在力促把现代技术应用到资源开发上去。同时还探索着种种新的途径,比如把制订规划和彼此有联系、但又互有差别的各类资源的有关研究结合起来。

2. 联合国秘书处正在为发展中国家派出的人员制订自然资源方面的训练计划。计划主要是经由区域性和跨区性的研讨会来实施的。这类研讨会也是一种把先进技术传授到发展中国家去的有效方法。研讨会有各种论题,比如与资源开发有关的制图学、海水淡化、能源政策,以及探矿中应用的地质化学技术等。

3. 联合国秘书长准备好了一份五年调查计划,打算采用经济上合算的、技术上先进的方法来对自然资源进行勘探和估价,从而为资源的开发做出贡献。该计划包括矿物资源、水资源、能源和电力方面的九点调查内容,分述如下:

(1) 矿物资源方面:调查发展中国家区域内近海大陆架的矿藏;调查世界铁矿石资源;调查重要的有色金属情况;调查发展中国家选定的矿山,要着眼于应用现代技术来找出矿物的新储藏,并提高其产量。

(2) 水资源方面:调查有可能缺水的那些发展中国家的需水情况和水资源;调查国际河流的开发潜力。

(3) 能源和电力方面:调查发展中国家潜在的地热能源;调查油页岩资源;调查发展中国家对小型发电设备的需求情况。

4. 上面所提出的九项调查,每一项都有两个目的:一是针对每个相关发展中国家的自然资源潜力,提供具有重大意义的新情报、新见解和新方法;二是搜集数据。这些数据,从该选定区域长期的潜在效力和需求来看,可能会提供一种全球视野。上述九项调查,在准备和选定项目时,对于提出多边或双边技术援助和财政援助的要求,也是有用的。

3 科技报道的文体特点及其翻译

科技报道,可分为一般性的科技报道和专业性较强的科技报道两类。

一般性的科技报道往往涉及某领域最新的进展,对专业知识涉及不深,使用的词汇也都是本专业的基础词汇。一般性的科技报道不以技术性的介绍为主,而是强调新技术带来的结果。因此,文章句子不长、结构简单、通俗易懂。翻译一般性的科技报道也要了解专业所用的基本术语。

专业性较强的科技报道刊登在某些学术期刊上,文体较正式,只有从事某一专业的人员才能看懂，翻译时需要较深厚的专业基础。在翻译过程中要注意专业词汇的翻译，同时还要注意译文的文体。

21st Century Refrigerator

Staring longingly into your empty refrigerator at 3 a.m. is about to become a whole lot cheaper. The US Department of Energy recently released new efficiency standards for refrigerators and freezers manufactured after July 1, 2002. To meet these standards, next generation refrigerators must use 30 percent less energy.

Manufacturers will achieve most of the energy savings through improvements in the compressor—lower-viscosity oil and tighter bearings, for example. Other changes will include thicker insulation, more efficient fans, and increased surface area of the condenser and evaporator coils. Manufacturers will also step up efforts to incorporate smart features, such as sensors that determine when defrosting is needed. "These are evolutionary, not revolutionary, changes," says Len Swatkowski, director of engineering for the Association of Home Appliance Manufacturers in Chicago.

The appliances will also likely employ a new blowing agent for foam insulation. The current agent—a hydrochlorofluorocarbon—will be banned as of January 1, 2003, because of its ozone-depleting properties. Tests of a potential replacement—a hydrofluorocarbon—show that it conveys heat about as well as the old chemical, with less harm to Earth's ozone layer.

If issues concerning efficiency, toxicity, and large-scale availability can be worked out, manufacturers will likely switch to the new blowing agent at the same time they incorporate the efficiency improvements.

The bottom line: Consumers will reap energy savings of about $20 per year for a typical 20-cubic foot refrigerator (the most common size). And although the new refrigerators will cost approximately $80 more, consumers will see significant savings over a refrigerator's average 19-year life. The new fridges will have the same features and usable space as conventional units.

The DOE has now turned its attention to air conditioners, ranges and ovens, clothes washers, water heaters, and fluorescent light ballasts. Within a few years, all will be subject to new energy standards.

分析：报道中使用了一些与冰箱有关的专业词汇，如 compressor、insulation、fan、condenser、evaporator coil、sensor、defrosting、foam insulation 等，翻译时要特别注意这些词的意思。

译文：
21 世纪的电冰箱

使用电冰箱的开销有望大大减低。美国能源部新近颁布文件对 2002 年 7 月 1 日以后生产的冰箱、冷柜规定了新的能耗标准。据此，下一代冰箱的能耗必须在现有基础上降低 30%。

对生产者来说，实现降低能耗的主要途径是改进压缩机性能，如降低机油黏度和缩小轴承耦合间隙。另外，还要加大密封层厚度，提高电风扇效率，扩大压缩机和散热线圈面积。生产者将加大革新力度，开发智能特色，如利用传感器确定最佳除霜时间等。芝加哥家用电器生产者联合会会长莱思·斯奥特考夫斯基指出：“这些改进是革新，而非革命。”

冰箱的泡沫密封膨化剂也将更新。目前使用的盐酸氟氯碳酸对臭氧层有破坏作用，规定于 2003 年 1 月 1 日后禁止使用。可能的替代品是氢氟碳酸。实验证明其导热性能与盐酸氟氯碳酸相同，但对地球臭氧层的破坏作用较小。

如果该产品能降低能耗、减少污染，同时批量供货不成问题的话，冰箱生产者很可能在落实降低能耗方案过程中启用这种新材料。

归根结底，冰箱革新的结果会给消费者带来好处。比如，使用普通 20 立方英尺冰箱的用户每年可省约 20 美元的电费开支。尽管新一代冰箱售价可能提高 80 美元左右，但冰箱的平均寿命为 19 年，纵观整个使用期，消费者还是从节能方面获益。革新后的冰箱将保留原有的特色和使用空间。

目前，能源部已将注意力转向空调器、灶具和炉具、洗衣机、热水器、荧光灯镇流器等产品的节能问题上。不出几年，也将对此类产品做出新的能耗标准。

The Midnight Sun

Researchers in Washington State are working on a generator that might replace internal-combustion engines in hybrid electric vehicles. It would give these vehicles a much greater driving range than solar-powered cars while nearly eliminating hydrocarbon and carbon monoxide emissions. And because the device operates at relatively low temperatures, it's not expected to produce oxides of nitrogen (NOx).

Unit Ten Civil Aviation

The patented device is a ten-inch-long generator dubbed the Midnight Sun. A burner inside the generator ignites a mixture of air and natural gas, heating a ceramic tube to more than 1,400℃; in effect, the tube becomes a tiny sun. Super-efficient but expensive gallium antimonide solar cells are arrayed around the glowing core to convert thermal energy into electricity that could power a car and charge its batteries. Because the natural-gas generator burns fuel continuously, carbon monoxide and hydrocarbon emissions would be one-fiftieth those of vehicles with internal-combustion engines.

The researchers have already built a prototype of the Midnight Sun that is powerful enough to operate a small television. The next step is to construct an eight-burner generator that will be installed in the university's Viking 23 vehicle. While the car could operate strictly on natural gas, it's intended to run primarily on battery power. The Midnight Sun generator is a backup system designed primarily for recharging the batteries and for long trips.

分析：本报道介绍了新产品"午夜太阳"，即一种发电机，翻译时要密切注意与发电机相关的专业词汇。

译文：

午夜太阳

华盛顿州的科研人员正在研制一种发电机，它可能取代现在用在混合式电动车中的内燃发动机。这种发电机能使电动车比太阳能汽车跑得远得多，还几乎不排放碳氢化合物和一氧化碳气体。此外，因为这种发电机的工作温度相对较低，所以也不会产生氮氧化物（NOx）。

这种 10 英寸长的专利，别名叫"午夜太阳"。发电机内部的燃烧室点燃空气和天然气的混合气，把陶瓷管加温到 1,400℃以上；实际上这时陶瓷管已成一颗小太阳。在这灼热的芯管周围排列着超高能的、但很贵的镓锑化合物太阳能电池，把热能转化为电能来驱动车，并给车上的电池充电。因为这种天然气发电机不间断地燃烧，所以一氧化碳和碳氢化合物的排放量只是使用内燃发动机车辆排放量的五十分之一。

研制者们已经制造出一个"午夜太阳"的样机，发出的电能足以供一台小电视机使用。下一步是制造具有八个燃烧室的发电机，并将其安装在该大学的 Viking 23 型车上。虽然这种车只能烧天然气，但是现在让它主要靠蓄电池电能来工作。"午夜太阳"发电机是一种辅助系统，主要用来为电池充电，供跑长距离使用。

4 实验指示与实验报告的文体特点及其翻译

为了证明自然科学的某条定理或某个规律，科研人员往往需要进行实验。对于实验如何进行以及如何得出结论有相对固定的文体格式。

语法上的特点是祈使句用得很多。使用祈使句是为了说明实验如何进行。句子一般较短，结构较简单。词汇上的特点是动词全部是常用动词。如果描述一项复杂的实验，会出现较专业的动词。在篇章结构方面，该文体有十分清楚的模式：实验如何进行的指示（第一段），实验结果的陈述（第二段），结论（第三段）。

下面我们不妨把实验指示和实验报告加以对照，从中可以看出它们的差异。

Turn a gas-jar upside down and burn a wooden splint under it for about a quarter of a minute. Close the jar with a cover and then put it the right way up on the bench. Next remove the cover and quickly add 2cm^3 of lime-water. Replace the cover and shake the jar.

分析：该实验指示全文使用祈使句，句子较短，多用主动语态。因为讲述的是一般情况，所以使用了一般现在时。

译文：倒放一个集气瓶，在其下燃烧一块薄木片，持续约15秒钟。用盖子盖住集气瓶，然后将它正放在实验桌上。下一步是取掉盖子，迅速加入2毫升石灰水。重新盖上盖子并摇动集气瓶。

First, a long glass tube was taken. The tube was closed at the top and was then completely filled with water. Next it was placed vertically in a large barrel half-full of water. When the bottom of the tube was opened, the water level in the tube fell to a height of approximately 10 meters above the water level in the barrel. As a result, a vacuum was left in the upper part of the tube. The water in the tube was supported by the atmospheric pressure. The height of the column of water could therefore be used to measure atmospheric pressure.

Unit Ten Civil Aviation

分析：该实验报告全文使用陈述句，句子较长，多用被动语态。因为讲的是过去的情况，所以使用过去时。

译文：首先取一根长玻璃管，将顶端封闭并盛满水，然后竖直地放在一只水半满的大桶中。当管的底部打开时，管中的水面只下降到大桶水面之上大约10米高度处。结果，在管的上部就会留下真空。管内的水为大气压力所支撑，因此水柱的高度可用来测量大气压力。

从以上两例我们可以看出实验指示和实验报告的不同，下面再看一例有关实验指示的翻译。

An Experiment

PRELIMINARY

The chemistry laboratory is a place where you will learn by observation what the behavior of matter is. Forget preconceived notions about what is supposed to happen in a particular experiment. Follow directions carefully, and see *what actually does happen*. Be meticulous (very exact and careful) in recording the true observation even though you "know" something else should happen. Ask yourself why the particular behavior was observed. Consult your instructor (teacher) if necessary. In this way, you will develop your ability for critical scientific observation.

EXPERIMENT I: DENSITY OF SOLIDS

The density of a substance is defined as its mass per unit volume. The most obvious way to determine the density of a solid is to weigh a sample of the solid and then find out the volume that the sample occupies. In this experiment, you will be supplied with variously shaped pieces of metal. You are asked to determine the density of each specimen and then, by comparison with a table of known densities, to identify the metal in each specimen. As shown in Table 1, density is a characteristic property.

Table 1　Densities of Some Common Metals, g/cc

Aluminum	2.7
Lead	11.4
Magnesium	1.8

Monel metal alloy	8.9
Steel (Fe, 1% C)	7.8
Tin	7.3
Wood's metal alloy	9.7
Zinc	7.1

PROCEDURE

Procure (obtain) an unknown specimen from your instructor. Weigh the sample accurately on an analytical balance.

Determine the volume of your specimen by measuring the appropriate dimensions. For example, for a cylindrical sample, measure the diameter and length of the cylinder. Calculate the volume of the sample.

Determine the volume of your specimen directly by carefully sliding the specimen into a graduated cylinder containing a known volume of water. Make sure that no air bubbles are trapped. Note the total volume of the water and specimen.

Repeat with another unknown as directed by your instructor.

QUESTIONS

1. Which of the two methods of finding the volume of the solid is more precise? Explain.

2. Indicate how each of the following affects your calculated density: (a) part of the specimen sticks out of the water; (b) an air bubble is trapped under the specimen in the graduated cylinder; (c) alcohol (density, 0.79g/cc.) is inadvertently substituted for water (density, 1.00g/cc) in the cylinder.

3. On the basis of the above experiment, devise a method for determining the density of a powdered solid.

4. Given a metal specimen from Table 1 in the shape of a right cone of altitude 3.5cm with a base of diameter 2.5cm. If its total weight is 41.82g, what is the metal?

实验

准备工作

化学实验室是你通过实验观察可以知道物质性状的地方。忘记一切在特定实验条件下可能会发生什么情况的预想，细心地按照指令观察事情发生的实况。在记录实地观察到的情况时、即使你"明知"会发生其他问题，

Unit Ten　Civil Aviation

也必须十分慎重，做到非常准确、极其仔细。要问一问自己，为什么会观察到这种特殊的情况。如有必要，向导师请教。只有这样，才会提高你具有批判性的科学观察能力。

试验一：固体的密度

物质密度的定义是单位体积的质量。确定某一固体密度的最简单方法是称出该固体样品的重量，再求出样品的体积。做这项实验时，你会得到形状不同的金属块。要求确定每一种金属样品的密度，再和已知密度的一张表对比，以识别每种样品是哪一种金属。表1所示密度是诸金属的特性。

表1　几种普通金属的密度（克/立方厘米）

铝	2.7
铅	11.4
镁	1.8
莫涅耳金属合金	8.9
钢（含百分之一碳的铁）	7.8
锡	7.3
伍德金属合金	9.7
锌	7.1

步骤

从导师处领取一块不明性质的样品，放到分析天平上准确地称出它的重量。

测出该样品适当部位的尺寸，以确定其体积。比如，对于一个圆柱体样品，要测出它的直径和长度，算出它的体积。

小心地把该样品放进盛有已知水量的量筒内，直接确定该样品的体积。要保证水里不含气泡。把样品和水加在一起的总体积记录下来。

遵照导师的指导，用另一块样品，重复上面步骤。

问题

1. 上述两种方法中，哪种求固体物的体积更准确？试说明之。

2. 指出下述各种情况怎样影响到你所计算出来的密度：（1）样品的一部分露出水面；（2）量桶内，样品下面隐有气泡；（3）量桶内错把酒精（密度为0.79克/立方厘米）当成了水（密度为1克/立方厘米）。

3. 根据上面的实验，想出一种确定粉末状固体物密度的方法。

4. 设表1中的一种金属样品，呈正圆锥状，高3.5厘米，底部直径2.5厘米，重41.82克，这是什么金属？

通过以上分析，我们不难看到，科技文章本身就包含了各种变体，每个变体有自己特定的文体特征。用某种变体的特点来覆盖其他变体的特色是不全面的。因此翻译科技文章要从实际出发，通过综合分析，最大限度地保持原文的风格。

Quiz

1. Translate the following scientific writings into Chinese.

    (1) Aerobatic Insects

    New technology allows scientists to study how an insect flies and the design of its wings in detail. This could help develop new aerodynamic technology. How does a dragonfly twirl and dive in flight? How do flies loop the loop, spin upside down and perform tight rolls? The aerobatics of insects are truly amazing, but what enables these creatures to excel in flight beyond any human flying machines?

    Modern technology is starting to give some answers. Examined by high-speed film and scanning electron microscopes insects have revealed that one secret of their success lies in the design of their wings.

    Insects have thin wings made of tough membranes. Birds and bats fly with the help of joints and muscles inside their wings. However, insects become and stay airborne with wings that are moved only with the body muscles. Their wings change shape subtly, due not only to muscle power but also to a combination of rigid and flexible veins allowing varying degrees of elasticity.

    (2) Light Radiation from Pulsed Discharges in Water

    Abstract: (*Aim*) To study the characteristics of plasma radiation from pulsed discharges in water. (*Methods*) Pyrometeroscilloscope system was used to measure the brightness of the plasma radiation. The spark channel temperature and the distributions of the irradiance as a function of wavelength were obtained through a grey-body radiation model. (**Results**) For a stored capacitor energy of 312.5J, the spark channel temperature of 5×10^4K was obtained with a peak current of 50KA. Most of the plasma radiation energy is distributed in the far ultraviolet region of the spectrum (10–200nm). (**Conclusion**) Radiation of plasma underwater corresponds to its electric characteristics. Ultraviolet radiation is one of the principal forms of energy which is released from a pulsed discharge in water.

(3) New Semiconductors

Scientists are working to develop new ways to make the semiconducting material used in computers and other modern electronic devices.

Semiconductors now are made from the element silicon. But scientists believe other substances would be more effective. They say the substances they are developing can carry electrical signals more rapidly than silicon. The substances give off light. This means tiny electrical systems could communicate with each other by light instead of through wires. And the new substances are not easily damaged by radiation or by high or low temperatures. This makes them useful in space satellites and rockets.

Engineers at Northwestern University near Chicago developed a semiconductor material called indium arsenide phosphide, a mixture of indium, phosphorus and arsenic. One of the engineers, Bruce Wessels, said computers using the substance could operate about three-to-six times faster than those using silicon. He also said it is easy to make the substance in a laboratory or factory.

The materials needed to make indium arsenide phosphide are costly—about one-hundred times more costly than silicon. But Professor Wessels said materials are only a small part of the total cost. And he expects computer systems using his material will be available within five years.

Another possible substitute for silicon is called gallium arsenide. It can carry electrical signals about five times faster than silicon. And it gives off light. However, gallium arsenide systems are costly. They easily break. And they keep too much of their heat.

Scientists at the University of Illinois recently said they have developed a way to use both gallium arsenide and silicon in the same tiny electrical system, or chip. The scientists said such a combined system would have the good effects of both materials. Like gallium arsenide, it would be fast. It would give off light. And it would resist radiation. Like silicon, it would not be costly. And it could be made large. The head of the Illinois team, Hadis Morkoc, said his technique has worked well on several transitor devices that control the flow of electrical currents.

Scientists at several other American universities and companies and at companies in Japan also are working on ways to use both substances in electrical systems.

Part B Reading

Civil Aviation Faces Green Challenge

As a fast-growing emitter of greenhouse gases, the aviation industry is under intense pressure to improve its fuel efficiency. Kurt Kleiner surveys its options.

Last Sunday, Boeing **rolled out** its latest airliner, the 787. It boasts 20% improvements in fuel efficiency, and is the first large airliner whose main structure and skin is built of light, carbon-fibre composites, instead of metal alloys. But the industry, under pressure over its role in global warming, is **scrambling** to improve efficiency yet further.

"Our goal is to take the industry to the next level, with respect to environmental responsibility", says Anthony Concil, a spokesman for the International Air Transport Association.

At last month's Paris Air Show, this effort took centre stage. Airbus called for global collaboration on the technical challenges ahead. Giovanni Bisignani, director-general of the International Air Transport Association, said there should be a "zero-emissions" plane by 2050.

Civil aviation accounts for about 3.5% of total greenhouse-gas emissions, according to the Intergovernmental Panel on Climate Change, but this could rise as air travel becomes more popular. And the pursuit of greater fuel efficiency is also being driven by airlines' fuel bills, given the high price of oil. "We know that in order to be able to market a new aircraft, it must be between 15% and 25% better in fuel consumption than the aircraft it's going to replace", says Rainer von Wrede, director of environmental affairs at Airbus.

One place that Boeing and Airbus, which between them build most of the world's large civilian airliners, look for such improvement is in aeroengines. Today's civilian jets use turbofans, in which a gas turbine at the rear of the engine drives a fan **mounted** just in front, which provides most of the **thrust**. A turbofan is quieter and more energy efficient than a pure gas-turbine engine, in which the thrust comes entirely from the jet of hot air shooting out the back. Efficiency will be increased if these fans are made bigger. But as they expand, so do the **ducts** that surround them, and their increased weight and **drag** soon cancel out the advantage of the bigger fan.

In an EcoJet proposal also **unveiled** at Paris, British budget airline EasyJet suggested a plane that would halve carbon dioxide emissions per passenger mile by 2015. It proposes using open **rotors**, in which the duct is eliminated. The bigger fans that this makes possible could create an engine that is 15% more efficient, says David Clarke, head of technology development at engine-maker Rolls-Royce in Derby, UK. But eliminating the duct increases noise and leaves less protection for the **cabin** in the event that a **blade** fails. EasyJet suggests positioning the turbofans above and behind the plane, where the tail and body would block the noise to the ground, and where a broken blade would miss the cabin.

A similar increase in efficiency could come from a geared turbofan. In a conventional turbofan, the gas turbine in the back of the engine turns the fan directly by means of a **shaft**. The fan is most efficient, though, at a much slower rotational speed than the most efficient speed for the turbine. Conventional engines are forced to operate at a **compromise** speed.

Pratt & Whitney, a jet-engine manufacturer based in Hartford, Connecticut, is already testing a new engine that places a **gearbox** between the turbine and the fan. It allows the fan to run at a third of the speed of the turbine, and will give a 15% increase in efficiency, says Bob Saia, head of next-generation products there.

Beyond the engines, airliner makers are looking to better body designs to improve fuel efficiency. The simplest improvement is to make existing body designs lighter. The Boeing 787, which is expected to fly commercially next year, is leading the way in this regard, using carbon-fibre composites for its entire **airframe**. Stronger composites, reinforced by carbon nanotubes, are widely regarded as the next step.

Unit Ten Civil Aviation

Engineers are also using computer simulation to further reduce drag from airliners. The wings on the latest version of the Boeing 737 generate 30% less drag than the ones they replaced, says Bill Glover, director of environmental performance strategy for Boeing in Seattle, Washington. Increasingly powerful **simulation** tools promise even better designs, adds Hans Weber, an aviation analyst with Tecop International in San Diego, California. This allows engineers to **vet** sophisticated new designs thoroughly on computer, before embarking on the much more costly business of building prototypes for tests in wind tunnels.

Additionally, manufacturers are **toying with** radical body concepts, such as the blended wing-body that was pioneered in the US Air Forces' B2 stealth bomber. In these designs the tube-like **fuselage** and tail are eliminated, turning the entire aircraft into a lift surface. A blended wing-body design currently being ground-tested by Boeing and NASA promises a 20% fuel reduction. But industry watchers are unsure if passengers would accept such a design: they'd have to sit in a wide, windowless compartment, and those near the wing ends would be shaken up and down whenever the airliner **banked**. Airport gates would also have to be entirely redesigned to accommodate such a plane.

The industry is also looking at how new fuels could help it to **assuage** environmental concerns. Biofuels have obvious attractions—provided the public accepts that their growth and combustion are generating no net carbon emissions. **Bioethanol**, the most popular biofuel for car engines, will freeze at high altitudes, but **biodiesel** and **biobutanol** could each be burned in conventional turbofan engines. Richard Branson, chairman of London-based Virgin Airlines, **pledged** to fly a commercial jet on biofuels in 2008, and is working with Boeing and the aeroengine division of General Electric to test **candidate** fuels.

The ultimate zero-emissions fuel may be proved to be **hydrogen**, which could easily be burned by conventional jet engines with no carbon emissions. First, the world will have to figure out how to make and distribute hydrogen cheaply and safely. And even given that, planes would require a redesign. Although liquid hydrogen provides three times the energy of the same weight of **kerosene**, it takes up seven times as much space. That would require aircraft with bigger fuel tanks, as well as the insulation to keep the liquid cool.

Once a plane exists with the cooling equipment that allows it to use liquid hydrogen as fuel, the potential exists for a more efficient all-electric plane, says Gerald Brown, a physicist at NASA Glenn Research Center in Cleveland, Ohio. Conventional generators are too heavy to be efficient on aircraft. But with a source of cold hydrogen, a superconducting generator and electric motors start to make sense, he argues, and might be more efficient than today's turbofans.

In the end, the aviation industry will seek to combine all of these approaches in a concerted bid to head off **accusations** that its rapid growth is **spurring** global warming. "We have a lot of creative people with a lot of great ideas. I'd say nothing's off the table", says Boeing's Glover.

New Words and Expressions

roll out 推出（新产品、服务等）
scramble 艰难（或仓促）地完成
mount 安装
thrust 推力
duct 管道
drag 阻力
unveil 揭晓
rotor 转子
cabin 机舱
blade 叶片；刀片
shaft 轴
compromise 折衷；妥协
gearbox 变速箱
airframe 机身，机体
simulation 模拟

vet 审查
toy with 不太认真地考虑
fuselage 机身；外壳
bank 倾斜飞行
assuage 缓和；减轻
bioethanol 生物乙醇
biodiesel 生物柴油
biobutanol 生物丁醇
pledge 保证；承诺
candidate 候选人
hydrogen 氢
kerosene 煤油
accusation 责备；谴责
spur 刺激；激励

Unit Ten Civil Aviation

📝 Exercises

1. Find out the English equivalents of the following Chinese terms from the passage.

   （1）航空业 （2）燃油效率
   （3）碳纤维复合材料 （4）全球变暖
   （5）国际航空运输协会 （6）巴黎航展
   （7）零排放 （8）温室气体排放
   （9）政府间气候变化专门委员会 （10）燃油费用
   （11）大型民航客机 （12）航空发动机
   （13）涡轮风扇 （14）燃气涡轮发动机
   （15）廉价航空公司 （16）开式转子
   （17）齿轮传动涡轮风扇 （18）转速
   （19）普惠公司 （20）碳纳米管
   （21）风洞 （22）隐形轰炸机
   （23）混合翼身设计 （24）净碳排放
   （25）超导发电机 （26）电动机

2. Translate the following sentences into Chinese.

   (1) It boasts 20% improvements in fuel efficiency, and is the first large airliner whose main structure and skin is built of light, carbon-fibre composites, instead of metal alloys.

   (2) One place that Boeing and Airbus, which between them build most of the world's large civilian airliners, look for such improvement is in aeroengines.

   (3) Today's civilian jets use turbofans, in which a gas turbine at the rear of the engine drives a fan mounted just in front, which provides most of the thrust.

   (4) EasyJet suggests positioning the turbofans above and behind the plane, where the tail and body would block the noise to the ground, and where a broken blade would miss the cabin.

(5) The Boeing 787, which is expected to fly commercially next year, is leading the way in this regard, using carbon-fibre composites for its entire airframe.

(6) This allows engineers to vet sophisticated new designs thoroughly on computer, before embarking on the much more costly business of building prototypes for tests in wind tunnels.

(7) A blended wing-body design currently being ground-tested by Boeing and NASA promises a 20% fuel reduction.

Part C Extended Reading

A Brief Introduction to the International Public Air Law

The enormous importance of air transport in today's world and in particular the increasing flow of air transport among the countries, and the economics and high technology developments give rise to a very natural curiosity of lawyers for our respective countries about the legal issues. The rapid increase of the volume of transport has resulted more in the increase of the flight figures of foreign and domestic aircraft. Therefore, in certain air routes, the frequency of the aircraft flights was already overcrowded.

As the volume of air passenger and **cargo carriage** on the global basis is increasing rapidly, and due to the technology advancement of aircraft operations, the world is becoming "a sphere of one day life". Accordingly, the airlines of many nations have been expanding their air routes overseas and have increased aircraft flights in order to induce more passengers and cargoes, and therefore have brought about a more serious competition among countries.

In particular, since the super-sonic aircraft and the air cargo container appeared, the volume of air passengers and cargo transport has been gradually increased year by year in the world. In the meantime, aircraft accidents have occurred more frequently while the operation of aircraft passenger and cargo have increased.

There are many aviation cases in the developed countries of Asia, Europe and America as well as developing countries. The characteristic features of the damages due to aircraft accidents have (1) the nature of great amounts of damage for compensation, (2) the nature of total loss (all or nothing), (3) the nature of instant (Augenblick), (4) the nature of the subordinateness relation to the ground (air traffic control system) and

(5) the nature of internationality. These aircraft accidents have **incurred** many disputes between the victims and the air carrier in deciding on the limits or unlimits of **liability** for compensation and the **appraisal** of damages.

Perhaps even more important, for the purpose of keeping their safety, orderly, rapid movement of aircraft and solving disputes between victims and air carriers, they must dutifully observe the **regulations**, administrative procedures, aeronautical acts and international conventions relating to the air transport.

The international air law relating to the transport of passengers and cargoes is based upon international agreements, **protocols** and conventions.

In the early days of international air transportation, there were no uniform rules of law governing the carriage of passengers or cargoes in the world.

The right and duty of passengers and cargo owners, and the carrier's liability within the domestic air transport have been regulated by the domestic aviation law or civil and commercial code in the civil law countries. Civilian aircraft of airlines and air carriers that provide services crossing international boundaries into the airspace of another country or countries are regulated to a greater or lesser degree by each national government involved.

The regulation of each flight by two or more **sovereign** states creates a need in most cases for some formal international agreement between states involved about how their commercial air services are to be carried out. States individually regulate numerous aspects of air services, such as licensing their own air carriers, requiring that **tariffs** and flight schedules be filled with their authorities, collecting statistics, protecting consumers, etc. On the other hand, some states involve themselves in regional regulations with group of states, although most common regulation of non-binding character. International civil aviation is legally governed by the principle of the sovereignty of a state over its airspace.

Unit Ten Civil Aviation

New Words and Expressions

cargo 货物
carriage 运输，运送
incur 招致；引发
liability 责任；义务
appraisal 鉴定；评价

regulation 法规；章程
protocol 协议，协定
sovereign 有主权的
tariff 关税

Exercises

1. **Find out the English equivalents of the following Chinese terms from the passage.**

   （1）航空运输   （2）法律问题
   （3）航线   （4）超音速飞机
   （5）航空货运集装箱   （6）全损
   （7）空中交通管制系统 （8）航空承运人
   （9）航空条例   （10）国际公约
   （11）国际航空法   （12）民用飞机
   （13）国际协议   （14）航空服务
   （15）航班时刻表   （16）领空

2. **Translate the following sentences into Chinese.**

   (1) The rapid increase of the volume of transport has resulted more in the increase of the flight figures of foreign and domestic aircraft.

   (2) As the volume of air passenger and cargo carriage on the global basis is increasing rapidly, and due to the technology advancement of aircraft operations, the world is becoming "a sphere of one day life".

   (3) The right and duty of passengers and cargo owners, and the carrier's liability within the domestic air transport have been regulated by the domestic aviation law or civil and commercial code in the civil law countries.

(4) Civilian aircraft of airlines and air carriers that provide services crossing international boundaries into the airspace of another country or countries are regulated to a greater or lesser degree by each national government involved.

(5) On the other hand, some states involve themselves in regional regulations with group of states, although most common regulation of non-binding character.

Unit Eleven
Computer Science

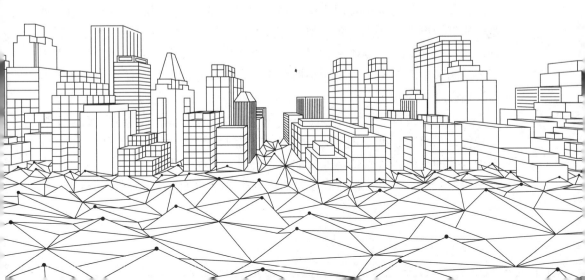

Part A Lecture

产品说明书的翻译

说明书文体包括各种产品的说明书、操作指南或使用手册、故障排除和维修保养方法等。从工业装置说明书,到一般机械或家用电器的使用说明,以及旅游指南、食品说明、医药说明和服务须知等均属此范围。

由于产品种类不同,说明书的内容及说明的方法也有所不同。机械说明书的内容一般包括产品特点、用途、规格、结构、性能、操作程序及注意事项等。家电产品说明书包括产品的安装方法、使用方法、常见问题的处理以及日常的维护与保养。食品说明书一般有食用方法、储藏方法、功能、构成成分、保质期等。医药说明书通常包括成分、适应症、用法、用量、注意事项、禁忌、副作用、有效期和储藏方式等。

尽管各类说明书由于构成不同,所以各有特点,但是,产品说明书的性质决定了各种说明书的共性。一般来说,产品说明书包括:产品的特征、功能和成分;安装使用/服用/饮用/食用的方式方法;注意事项;主要性能指标及规格。

由于说明书文体旨在介绍使用方法、操作方式或注意事项等,因此它的文体具有简略性、技术性、描述性和客观性的特点:(1)简略性表现为内容简略、文字浅显、句式简单,大量使用祈使句和省略句,避免使用不必要的修辞手段。(2)技术性表现为说明书中的文字涉及某方面的专业知识,即使是一般家用电器的使用说明书也具有技术性,工用机械说明书、医药说明书则技术性更强。(3)描述性表现为文字具有程序性和说服力。(4)客观性表现为说明书中的词语恰当、如实地反映产品的真实情况,旨在让消费者了解产品的性能、特点等。

从以上分析可知,产品说明书文体与其他文体不同,翻译时,需要掌握其特征,用符合说明书的语言如实地将产品说明书的内容翻译出来。首先,必须选择准确的语言来传递原文的真正信息。其次,需要了解产品的基本原理等相关的专业知识。遇到带有文学性语言的说明书时,还需要把握好分寸,

Unit Eleven　Computer Science

既要在译文中再现原文的文学色彩，又不能夸大其词。下面对各类产品的说明书进行分析和翻译。

1 一般机械或家用电器说明书的翻译

一般机械或家用电器说明书涉及产品的性能、结构、使用方法、操作方法及保养维修等方面内容。以下通过洗衣机、台灯、计算机使用说明书来分析此类说明书的翻译方法。

1·1 洗衣机使用说明书的翻译

Many parts of this machine are made of flammable plastic. Never place hot or burning objects on or near the washing machine.

When disconnecting the power cord from the power outlet, always take hold of the plug, and not the wire, and pull free. Never connect or disconnect the power plug with wet hands since you may receive an electric shock.

For really dirty clothing use hot water 40℃.

For removal of blood stains use cold water only.

STAIN REMOVAL AND BLEACHING

- Add 1/2 cap per liter of water.
- Soak laundry well in solution for at least 20–30 minutes and wash.
- Rinse thoroughly.

Power source: 220V/50Hz

Power consumption: 400W

Washing capacity: 3kg

Spin capacity: 3kg

Water supply pressure: 0.3kg/cm^2–10kg/cm^2

Net weight: 30kg

Dimension: 500mm×500mm×850mm

239

分析：以上洗衣机使用说明书客观地描述了洗衣机的性能、规格等有关情况，文中没有任何渲染、劝说性的语言。翻译时，应该用客观的词语来说明洗衣机的使用方法、性能及规格等。

译文：本洗衣机有许多可燃性塑料零部件，切勿在洗衣机上或附近放置热的或正在燃烧的物品。

从电源插座拔出电源插头时，不要拽电线，务必拿住插头，然后拔出。切勿用湿手插上或拔出电源插头，以防触电。

清洗过脏衣物请用 40℃热水。

清洗血渍，请用冷水。

漂洗斑渍和漂白

- 每千克水用 1/2 瓶盖漂剂。
- 至少用 20 分钟—30 分钟时间彻底浸泡衣物，然后再洗干净。
- 彻底漂洗干净。

　　电源：220V/50Hz

　　耗电量：400W

　　洗涤容量：3kg

　　脱水容量：3kg

　　使用水压：0.3kg/cm^2–10kg/cm^2

　　净重：30kg

　　外形尺寸：500mm × 500mm × 850mm

1·2 台灯使用说明书的翻译

Thank you for your purchase of a Philips desk light!

Caution Warnings

For your protection, please read this manual and the safety warnings carefully before using the desk light and keep it for later reference. The manufacturer will not be held responsible for any damages caused by improper usage or modifications to the desk light.

Electrical Specifications

Unit Eleven Computer Science

Voltage/Frequency:	220V/50Hz
Power/Lamp:	27W/PL-F27W/4P
Lampholder:	GX 10q-4
Electrical Insulation Classification:	II

Operation Instructions

This desk light is not waterproof and is only suitable for indoor usage.

No alterations of any kind should be carried out on this desk light.

Do not use any voltage exceeding a 10% margin of the specified standard.

Do not touch the lamp inside or the lamp shade when the power is switched on.

Do not place flammable material near the desk light.

In case of operation failure, please switch off the desk light, unplug the power cord and contact your nearest Philips dealer.

This desk light is suitable for Philips PL-F/4P 27W compact fluorescent lamps only.

Installation

Take out the desk light body, base and clip.

To use the desk light with the base, insert the stem onto the base plate and tighten the fixing screw.

To use the desk light with the clip, mount the clip onto the desired surface and tighten the fixing screw, then insert the stem onto the clip.

The supplied clip is not suitable for use on tubes.

Maintenance

In order to ensure optimum performance, we recommend you to clean the desk light twice a year. When cleaning the desk light take care to use soft cotton cloths only.

Do not use chemical solvents to clean the desk light, it might damage the painted surfaces.

Caution: Please turn off the power and unplug the desk light before replacing the lamp!

分析：以上台灯说明书描述了台灯的规格、操作和使用方法、维护和保养等有关情况，语言简洁、层次分明。译义也应体现原文的特点。

译文：

感谢您购买飞利浦台灯！

注意事项

为了您的安全，请在使用灯具前阅读本说明书，并保留本说明书作日后参考。所有不当使用或自行更改而导致的产品损坏，生产商不负任何责任。

灯具电气规格

电压/频率：	220V/50Hz
功率/灯管：	27W/PL-F 27W/4P
灯座：	GX 10q-4
电气绝缘等级：	II

操作说明

此台灯不防水，只适合在室内使用。

不可随意更改产品结构。

所有电源勿超过额定电压的 10%。

台灯通电时，请勿触摸光源和灯罩。

请勿将易燃物品放在灯具附近。

如发生故障，请关掉台灯，拔下电源线，并与最近的飞利浦经销商联系。

此产品只适用飞利浦 PL-F/4P 27W 电子节能灯管。

安装步骤

取出灯具、底座、灯夹。

使用底座：将灯杆插入底座安装孔，拧紧固定螺丝。

使用灯夹：将灯夹固定在安装面的边缘，拧紧固定螺丝，灯杆插入灯夹的安装孔。

所提供灯夹不适合安装在圆管上。

保养方法

为了保持灯具明亮效果，请定期（半年一次）进行清洁。清洁灯具时，只需使用柔软的棉布即可。

请勿使用化学溶剂清洗灯具，否则易损坏油漆表面。

注意：更换光源时，请务必先关闭电源！

Unit Eleven   Computer Science

1·3 计算机使用说明书的翻译

CAUTION: Safety Instructions

Use the following safety guidelines to help ensure your own personal safety and to help protect your computer and working environment from potential damage.

SAFETY: General Safety

Observe the following safe-handling guidelines to ensure personal safety:

- When setting up the computer for work, place it on a level surface.

- Do not attempt to service the computer yourself, except as explained in your Dell™ documentation or in instructions otherwise provided to you by Dell. Always follow installation and service instructions closely.

- To help avoid the potential hazard of electric shock, do not connect or disconnect any cables or perform maintenance or reconfiguration of this product during an electrical storm. Do not use your computer during an electrical storm.

- Do not push any objects into the air vents or openings of your computer. Doing so can cause fire or electric shock by shorting out interior components.

- If your computer includes a modem, the cable used with the modem should be manufactured with a minimum wire size of 26 American wire gauge (AWG) and an FCC-compliant RJ-11 modular plug.

- If your computer has both a modem RJ-11 connector and a network RJ-45 connector, which look alike, make sure that insert the telephone cable into the RJ-11 connector, not the RJ-45 connector.

- Keep your computer away from radiators and heat sources. Also, do not block cooling vents. Avoid placing loose papers underneath your computer; do not place your computer in a closed-in wall unit or on a bed, sofa, or rug.

- Do not use your computer in a wet environment, for example, near a bath tub, sink, or swimming pool or in a wet basement.

- Do not spill food or liquids on your computer.
- Before you clean your computer, disconnect the computer from the electrical outlet. Clean your computer with a soft cloth dampened with water. Do not use liquid or aerosol cleaners, which may contain flammable substances. Allow the computer to dry before reconnecting the power cord to the electrical outlet.
- Ensure that nothing rests on your computer's cables and that the cables are not located where they can be stepped on or tripped over.

CAUTION

- Do not operate your computer with any cover(s) (including computer covers, bezels, filler brackets, front-panel inserts, and so on) removed.
- PC Cards may become very warm during normal operation. Use care when removing PC Cards after their continuous operation.

WARNING

The cord on this product contains lead, a chemical known to the State of California to cause birth defects or other reproductive harm. *Wash hands after handling.*

分析：以上计算机使用说明书着重描述了计算机的安全操作方法，词汇通俗易懂，句子结构简单。文中出现了一些与计算机相关的专业词汇，翻译前应先了解一些计算机方面的知识，如计算机各部件的名称等。

译文：

警告：安全说明

请遵循以下安全原则，以确保您个人的人身安全，并且避免您的计算机和工作环境受到潜在的损坏。

安全：常规安全

为确保个人安全，请遵循以下安全操作指南：

- 启动计算机工作时，请将其置于水平面上。
- 除了在 Dell™ 说明文件中或在 Dell 另外提供的说明中阐明之外，请勿自行拆装计算机。请始终严格遵循安装与维修说明。
- 为避免电击的潜在危险，请勿在雷雨期间连接或断开任何电缆，也请勿在此期间维修或重新配置本产品。雷雨期间，请勿使用本产品。
- 请勿将任何物体塞入计算机的通风孔或开口处。否则会导致内部组件短路，从而引起火灾或遭到电击。

Unit Eleven　Computer Science

- 若计算机安装了调制解调器，请使用美国线规（AWG）规定的 26 号电缆以及符合美国联邦电信委员会（FCC）标准的 RJ-11 通讯网路用接头。
- 若计算机既有调制解调器 RJ-11 连接器又有网络 RJ-45 连接器，请区分两种接口，确保将电话线插入 RJ-11 连接器，而不是插入 RJ-45 连接器中。
- 使计算机远离暖气片和热源。另外，请勿堵塞冷却通风孔。请勿将松散的纸张放置到计算机下面；请勿将计算机置于封闭的壁柜内或置于床、沙发或小地毯上。
- 请勿在潮湿的环境如浴缸、洗碗池、游泳池旁边或潮湿的地下室内使用计算机。
- 请勿将食物或液体洒落到计算机上。
- 清洁计算机之前，请先断开计算机与电源插座的连接。使用蘸水的软布擦拭计算机。请勿使用液体或喷雾清洁剂，因为其中可能含有易燃物质。请在计算机干燥之后再将电源线连接到电源插座上。
- 请确保没有物品压住计算机的电缆，并将电缆置于不易被人踩踏或不易绊脚的地方。

注意

- 若任何盖板（包括主机盖、挡板、填充挡片、前面板插件等）被移除，请勿操作计算机。
- 正常运行期间，PC 卡可能会变热。在 PC 卡连续运行后将其卸下时，请务必小心。

警告

　　该产品上的电线含铅，在加利福尼亚州普遍认为，铅是一种已知会造成先天缺陷或其他生殖损害的化学物质。**在接触电线后请洗手。**

 医药说明书的翻译

医药说明书涉及专业知识，具有很强的技术性。请看下面一例。

例 4

ISOKET

For long-term and/or emergency treatment of angina pectoris.

Composition:

Isosorbide dinitrate 20 mg per tablet, with prolonged action.

245

Indications:

Angina pectoris, especially the so-called "effort angina"; prevention and subsequent treatment of myocardial infarction. Angina pectoris vasomotorica, a form induced by central nervous disorder, is not affected.

Contra-indication:

Glaucoma.

Dosage:

One tablet of Isoket retard to be taken with water and without chewing in the morning after breakfast, and one in the evening before retiring; this will protect the patient from attacks of angina pectoris during both the day and the night. If an attack does occur, half a tablet should be chewed and then swallowed, whereby an immediate action is produced.

Note:

No side-effects are to be expected, but some patients may develop the well-known "nitrate headache" which in itself is harmless. In such instances, it is recommended that the dosage be reduced. Because this nitrate effect is subject to tachyphylaxis, the Isoket retard medication need not be interrupted. Security for angina pectoris patients for 8–10 hours. In cases of extreme severity, immediate relief can be obtained by chewing Isoket retard.

Packings of 60 tablets.

Made in Germany

分析：技术性文章的翻译要求准确，如有差错，会引起严重后果。在翻译医药说明书时，首先要注意一些术语的翻译，如 composition（成分）、description（性状）、action（作用）、indication（适应症）、contra-indication（禁忌症）、precaution（注意事项）、side effects（副作用）、dosage and administration（剂量和用法）、packing（包装）、expiry date（失效日期）、manufacturing date（出厂日期）等。其次，对药品名称的翻译也要慎重，如例4，可按音译将"Isoket"译为"异速凯他"；为了使中国读者更明白药品的化学性质，也可将"Isoket"译为"硝酸异山梨酯"。最后，翻译时也要注意某些疾病名称的翻译，如 angina pectoris（心绞痛）、myocardial infarction（心肌梗死）、central nervous disorder（中枢神经紊乱）、angina pectoris vasomotorica（血管舒缩性心绞痛）、glaucoma（青光眼）、nitrate headache（硝酸盐性头痛）等。

Unit Eleven　Computer Science

译文：　　　　　　　　异速凯他（硝酸异山梨酯）

供心绞痛长期或急救使用。

【成分】每片含二硝酸异山梨醇酯 20 毫克。长效。

【适应症】专治心绞痛，特别是劳力性心绞痛。预防和治疗心肌梗塞。但对由中枢神经紊乱而引起的血管舒缩性心绞痛无作用。

【禁忌症】青光眼。

【剂量】早饭后用开水服 1 片，不用咀嚼。晚间睡前服 1 片，可避免患者在白天和晚间心绞痛发作。如心绞痛已发作，应咀嚼半片，咽下后即可立刻发挥作用。

【注意事项】一般不会发生副作用。有些患者可能出现常见的、并无大害的"硝酸盐性头痛"。如遇此种情况，可酌减剂量。因为患者对这种硝酸盐作用易产生快速耐受性，所以不必停药。本品可在 8 小时至 10 小时内保护心绞痛患者安全。严重患者咀嚼本品可立即缓解。

【包装】60 片包装。

德国制造

3　食品说明书的翻译

食品说明书涉及某种食品的食用方法、功能、构成成分等，请看下面例子。

Lacovo

Lacovo, scientifically prepared with choice ingredients including malt extract, milk, powder cocoa, fresh butter and eggs, is rich in vitamins A, B, D and organic phosphorous. It promotes health and aids convalescence and is especially good for neurasthenia and mental exhaustion. Take regularly, Lacovo helps build up body resistance against disease. A nourishing beverage for all ages. An excellent gift in all seasons.

For drinking hot: Put two or three teaspoonfuls of Lacovo in a cup, then add hot water and stir until the grains are thoroughly dissolved. Add sugar and milk to taste.

For drinking cold: Put two or three teaspoonfuls of Lacovo in a glass of cold water and stir until the grains are thoroughly dissolved. Then add fresh milk or condensed milk. It makes a delightful and wholesome drink in summer.

分析：以上是一种饮品的说明书。说明书的第一句话客观地描写了产品的构成情况，而后面的句子出现了一些具有文学色彩的渲染性语言。翻译时，需将原文中不同色彩的语言在译文中再现，以达到语义信息和风格信息的对等。

译文： 乐口福

　　乐口福系采用麦精、牛乳、可可粉、鲜黄油、鸡蛋等原料，以科学方法配制而成，富含维生素 A、B、D 及天然有机磷质，故能增强体质、恢复健康，对用脑过度、神经衰弱更有裨益。常饮本品，尤能增强抗病能力。本品营养丰富、老少皆宜，且为馈赠亲友之四季佳品。

　　热饮：取乐口福二至三茶匙放在杯中，然后注入适量开水，搅拌至完全融化。加入适量糖或牛奶，口感更佳。

　　冷饮：先将冷水注入杯中，然后加入乐口福二至三茶匙，搅拌至完全融化，加入适量鲜奶或炼乳，口感更佳。本品诚为夏季消暑保健之饮品。

1. Translate the following product specifications into Chinese.

(1) Nutrition Information

　　B-Complex is a combination of B-Vitamins, which helps convert food into energy and maintain normal nervous system function. This product contains 100% of the US Reference Daily Intake (RDI) of folate. Studies show folate helps men and women maintain normal cardiovascular system function. Adequate folate in healthful diets may reduce a woman's risk of having a child with brain or spinal cord birth defects.

　　Each tablet in this bottle is clear coated for easy swallowing.

(2) Warnings on Using Tapes

　　Panasonic VHS Video Cassette

　　Exclusively to be used with VHS Video Recorder

Unit Eleven　Computer Science

HOW TO USE AND STORE

To Use

- To prevent accidental erasure of recorded material, the VHS cassette has a removable tab on the rear side.
- The tab can be broken off by a small screwdriver. After it has been removed, the tape can be used for playback only.
- To record on a cassette from which the tab has been removed, simply cover the hole with a piece of adhesive tape.

To store

- Always put the cassette back in its case before storage. Store in a vertical position.
- Keep in a cool and dry place and avoid storage in direct sunlight or near sources of heat.
- Do not drop or otherwise subject the cassette to shock impact.
- For best results, store the cassette with the tape fully rewound.
- If the tape has been rewound unevenly, rewind it again to pack the tape properly.
- The tape in a VHS cassette cannot be spliced. Do not attempt to repair or open the cassette.
- Do not touch the tape surface itself. Dirt and oil from your skin may deteriorate the tape coating.

(3)　Directions for TV Maintenance

- The set is highly sensitive and should be handled with care.
- Don't install the set in a location near heat sources, such as radiators and stoves, or in a place subject to excessive dust, mechanical vibration or shock.
- Avoid placement and operation in direct sunlight or in very humid places.

- Take care to allow adequate air circulation on all sides to prevent internal heat build-up.
- Don't place the set on soft surfaces (rugs, blankets, etc.) or near curtains or draperies, as they may block the ventilation holes.
- Don't open the cabinet, inside which are present dangerously high voltages.
- Save the shipping carton and packing material for the set, which will come handy if it has to be repacked and reshipped. Repack the set as it was originally done at the factory.
- Clean the set periodically with a soft cloth so as to keep it looking always new.
- Stubborn stains may be removed with a cloth slightly dampened with a little mild detergent solution. Never use strong solvents, such as thinner or benzene, or abrasive cleansers, as they will damage the cabinet.
- Unplug the set from the wall outlet if it is not to be used for a long period of time. Pull the mains lead-in by the plug when disconnecting it. Never pull the mains lead-in itself.

Part B Reading

Artificial Intelligence

Artificial intelligence (AI) seeks to make computers do the sorts of things that minds can do.

Some of these (e.g., reasoning) are normally described as "intelligent". Others (e.g., vision) aren't. But all involve psychological skills—such as **perception**, association, prediction, planning, motor control—that enable humans and animals to attain their goals.

Intelligence isn't a single dimension, but a richly structured space of diverse information-processing capacities. Accordingly, AI uses many different techniques, **addressing** many different tasks.

And it's everywhere.

AI's practical applications are found in the home, the car (and the driverless car), the office, the bank, the hospital, the sky and the Internet, including the Internet of Things (which connects the ever-multiplying physical sensors in our **gadgets**, clothes, and environments). Some lie outside our planet: robots sent to the Moon and Mars, or satellites orbiting in space. Hollywood **animations**, video and computer games, sat-nav systems, and Google's search engine are all based on AI techniques. So are the systems used by financiers to predict movements on the stock market, and by national governments to help guide policy decisions in health and transport. So are the apps on mobile phones. Add **avatars** in virtual reality, and the toe-in-the-water models of emotion developed for "companion" robots. Even art galleries use AI—on their websites, and also in exhibitions of computer art. Less happily, military drones roam today's battlefields—but, thankfully, robot minesweepers do so too.

AI has two main aims. One is technological: using computers to get useful things done (sometimes by employing methods very unlike those used by minds). The other is scientific: using AI concepts and models to help answer questions about human beings and other living things. Most AI workers focus on only one of these, but some consider both.

Besides providing countless technological **gizmos**, AI has deeply influenced the life sciences.

In particular, AI has enabled psychologists and neuroscientists to develop powerful theories of the mind–brain. These include models of how the physical brain works, and—a different, but equally important, question—just what it is that the brain is doing: what computational (psychological) questions it is answering, and what sorts of information processing enable it to do that. Many unanswered questions remain, for AI itself has taught us that our minds are very much richer than psychologists had previously imagined.

Biologists, too, have used AI—in the form of "artificial life" (A-Life), which develops computer models of differing aspects of living organisms. This helps them to explain various types of animal behaviour, the development of bodily form, biological evolution, and the nature of life itself.

Besides affecting the life sciences, AI has influenced philosophy. Many philosophers today base their accounts of mind on AI concepts. They use these to address, for instance, the **notorious** mind–body problem, the **conundrum** of free will, and the many puzzles regarding consciousness. However, these philosophical ideas are hugely **controversial**. And there are deep disagreements about whether any AI system could possess real intelligence, creativity, or life.

Last, but not least, AI has challenged the ways in which we think about humanity and its future. Indeed, some people worry about whether we actually have a future, because they foresee AI surpassing human intelligence **across the board**. Although a few thinkers welcome this prospect, most **dread** it: what place will remain, they ask, for human dignity and responsibility?

Unit Eleven Computer Science

New Words and Expressions

perception 感知
address 设法解决；处理
gadget 小装置；小器具
animation 动画
avatar 化身
gizmo 小玩意儿；小发明
notorious 众人皆知的；臭名昭著的
conundrum 难题
controversial 有争议的
across the board 全面地
dread 恐惧；担心

Exercises

1. **Find out the English equivalents of the following Chinese terms from the passage.**

   （1）运动控制　　　　　　　（2）无人驾驶汽车
   （3）物联网　　　　　　　　（4）传感器
   （5）卫星导航系统　　　　　（6）搜索引擎
   （7）股市走势　　　　　　　（8）虚拟现实
   （9）军用无人机　　　　　　（10）扫雷机器人
   （11）生命科学　　　　　　（12）生物进化
   （13）人类智慧

2. **Translate the following sentences into Chinese.**

   (1) But all involve psychological skills—such as perception, association, prediction, planning, motor control—that enable humans and animals to attain their goals.

   (2) So are the systems used by financiers to predict movements on the stock market, and by national governments to help guide policy decisions in health and transport.

   (3) Less happily, military drones roam today's battlefields—but, thankfully, robot minesweepers do so too.

   (4) Many unanswered questions remain, for AI itself has taught us that

our minds are very much richer than psychologists had previously imagined.

(5) They use these to address, for instance, the notorious mind-body problem, the conundrum of free will, and the many puzzles regarding consciousness.

(6) Last, but not least, AI has challenged the ways in which we think about humanity and its future. Indeed, some people worry about whether we actually have a future, because they foresee AI surpassing human intelligence across the board.

Part C Extended Reading

Big Data

Where does all the data in "big data" come from? And why isn't big data just a concern for companies such as Facebook and Google? The answer is that the web companies are the forerunners. Driven by social, mobile, and cloud technology, there is an important transition taking place, leading us all to the data-enabled world that those companies **inhabit** today.

From Exoskeleton to Nervous System

Until a few years ago, the main function of computer systems in society, and business in particular, was as a digital support system. Applications digitized existing real-world processes, such as word-processing, payroll, and inventory. These systems had **interfaces** back out to the real world through stores, people, telephone, shipping, and so on. The now-quaint phrase "paperless office" **alludes** to this transfer of pre-existing paper processes into the computer. These computer systems formed a digital exoskeleton, supporting businesses in the real world.

The arrival of the Internet and the Web has added a new dimension, bringing in an era of entirely digital business. Customer interaction, payments, and often product delivery can exist entirely within computer systems. Data doesn't just stay inside the exoskeleton any more, but is a key element in the operation. We're in an era where business and society are acquiring a digital nervous system.

An organization with a digital nervous system is characterized by a large number of inflows and outflows of data, a high level of networking, both internally and externally, increased data flow, and consequent complexity.

This transition is why big data is important. Techniques developed to deal with interlinked, heterogeneous data acquired by massive web companies will be our main tools as the rest of us experience the transition to the digital native operation. We see early examples of this, from catching **fraud** in financial transactions to **debugging** and improving the hiring process in HR: and almost everybody already pays attention to the massive flow of social network information concerning them.

Charting the Transition

As technology has progressed within business, each step taken has resulted in a leap in data volume. To people looking at big data now, a reasonable question is to ask why, when their business isn't Google or Facebook, does big data apply to them?

The answer lies in the ability of web businesses to conduct 100% of their activities online. Their digital nervous system easily stretches from the beginning to the end of their operations. If you have factories, shops, and other parts of the real world within your business, you've further to go in incorporating them into the digital nervous system.

But "further to go" doesn't mean it won't happen. The drive of the Web, social media, mobile, and the cloud is bringing more of each business into a data-driven world. In the UK, the Government Digital Service is unifying the delivery of services to citizens. The results are a radical improvement of citizen experience, and for the first time many departments are able to get a real picture of how they're doing. For any **retailer**, companies such as Square, American Express, and Foursquare are bringing payments into a social, responsive data ecosystem, liberating that information from the **silos** of corporate accounting.

What does it mean to have a digital nervous system? The key trait is to make an organization's feedback loop entirely digital. That is, a direct connection from sensing and monitoring inputs through to product outputs. That's **straightforward** on the Web. It's getting increasingly easier in retail. Perhaps the biggest shifts in our world will come as sensors and robotics bring the advantages web companies have now to domains such as industry, transport, and military.

The reach of the digital nervous system has grown steadily over the past 30 years, and each step brings gains in **agility** and flexibility, along with an order of magnitude more data. First, from specific application programs to general business use with the PC. Then, direct interaction over the Web. Mobile adds awareness of time and place, along with instant notification. The next step, to cloud, breaks down data silos and adds storage and compute elasticity through cloud computing. Now, we're integrating smart agents, making them act on our behalf and connect to the real world through sensors and automation.

Coming, Ready or Not

If you're not **contemplating** the advantages of taking more of your operation digital, you can bet your competitors are. As Marc Andreessen wrote last year: "Software is eating the world." Everything is becoming programmable.

It's this growth of the digital nervous system that makes the techniques and tools of big data relevant to us today. The challenges of massive data flows, and the erosion of hierarchy and boundaries, will lead us to the statistical approaches, systems thinking, and machine learning we need to cope with the future we're inventing.

New Words and Expressions

inhabit 居住于；占据
exoskeleton 外骨骼
interface 界面；接口
allude 暗指
fraud 欺诈罪；骗子
debug 排除故障；调试（程序）
retailer 零售商
silo 筒仓；孤岛
straightforward 简单的；坦诚的
agility 敏捷性
contemplate 沉思；思量

Exercises

1. Find out the English equivalents of the following Chinese terms from the passage.

（1）云技术　　　　　　　　（2）神经系统
（3）文字处理　　　　　　　（4）工资单
（5）库存　　　　　　　　　（6）无纸化办公
（7）客户交互　　　　　　　（8）产品交付
（9）数据流入　　　　　　　（10）数据流出
（11）数据流　　　　　　　　（12）金融交易
（13）数据量　　　　　　　　（14）社交媒体
（15）数据生态系统　　　　　（16）反馈回路
（17）即时通知　　　　　　　（18）数据孤岛
（19）云计算　　　　　　　　（20）智能代理
（21）系统思维　　　　　　　（22）机器学习

2. Translate the following sentences into Chinese.

(1) Driven by social, mobile, and cloud technology, there is an important transition taking place, leading us all to the data-enabled world that those companies inhabit today.

(2) An organization with a digital nervous system is characterized by a large number of inflows and outflows of data, a high level of networking, both internally and externally, increased data flow, and consequent complexity.

(3) Techniques developed to deal with interlinked, heterogeneous data acquired by massive web companies will be our main tools as the rest of us experience the transition to the digital native operation.

(4) We see early examples of this, from catching fraud in financial transactions to debugging and improving the hiring process in HR: and almost everybody already pays attention to the massive flow of social network information concerning them.

(5) Perhaps the biggest shifts in our world will come as sensors and robotics bring the advantages web companies have now to domains such as industry, transport, and military.

(6) The challenges of massive data flows, and the erosion of hierarchy and boundaries, will lead us to the statistical approaches, systems thinking, and machine learning we need to cope with the future we're inventing.

Unit Twelve
Mechanical Engineering

Part A Lecture

科技文章检索

科技文献是用文字、图形、符号、音频、视频等手段记录科技信息或知识的载体。科技文献资料包括图书、科技期刊、科技报告、专利文献、科技会议文献、学位论文、标准文献、产品资料、政府出版物、科技档案、报纸等。

科技文献检索是一项独特的技能。在"知识爆炸""信息爆炸"的今天，大学生不仅要掌握本专业所需的基础理论和专业知识，还要重视科技文献检索能力的培养与训练。据说在德国柏林图书馆的大门上刻着这样一行字："这里是人类知识的宝库，如果你掌握了它的钥匙，那么这里的全部知识都是你的。"这里所说的"钥匙"就是文献资料检索能力。掌握和运用检索文献资料的钥匙，及时有效地查阅到所需的学习和研究资料的方法，是我们都必须具备的基本功。

怎样才能检索到我们在学习和研究中所需的文献资料？下面我们简单介绍文献资料的检索，然后谈谈在学习和研究中如何利用图书馆馆藏文献、数据库以及网络等资源检索文献资料。

1 文献资料检索概述

我们在查阅文献资料的过程中，知道所要检索的课题要求后，就要选定检索工具，确定检索的途径和方法；在查到一批资料以后，还要进行分析研究，去粗取精，去伪存真。

1.1 检索文献资料的主要工具

检索工具是人们用以积累和查找文献线索的手段。它具有存储和检索两个功能。一方面，它能将有关文献的特点著录下来，形成一条条的文献线索，并将它们按一定的方法排列起来，以供检索，这就是文献的存储过程；另一方面，它又能提供一定的检索途径，使人们按照一定的检索方法查出所需要的文献，这就是文献的检索过程。

Unit Twelve　Mechanical Engineering

为了查找和利用各种文献资料，我们必须了解检索工具的类型，熟悉检索工具的出版形式，掌握国内外一些常用检索工具的特点、作用和使用方法。

（1）检索工具的类型

检索工具的种类繁多，从检索方式划分，可分为手工检索工具和计算机检索工具；从出版形式划分，可分为书本式、卡片式和计算机可读式；从编著方式划分，可分为目录、题录、文摘和索引。

（2）检索工具的内容结构

检索工具的内容大致由下列四部分组成：

A. 著录部分

　　检索工具存储的不是文献的全文，而是描述文献外表特征以及内容特征的著录。著录部分是检索工具内容的主题结构。著录部分是否能起到检索作用，关键在于著录部分的组织编排方式。著录部分的组织编排方法有分类、主题、书刊名称、著者姓名、文献序号等。

B. 索引部分

　　著录部分的分类编排虽然可以起到一定的检索作用，但检索效率很低，而通过分类索引、主题索引、著者索引、号码索引等方法，能迅速、准确、全面地查阅到所需文献资料。索引是检索工具检索功能的标志，打开检索工具存储知识宝藏的钥匙就是索引。

C. 说明部分

　　说明部分的内容一般包括编制目的、适用范围、收录年限、著录款目、查阅例述、注意事项等。

D. 附录部分

　　附录部分是检索工具内容的必要补充。其内容包括收录的文献类型范围、摘用的期刊种类、刊名的缩写、核心期刊的目录、不同文字转译的对照表、缩写术语的解释等。

1·2 检索文献资料的基本途径

查找文献资料要根据文献的不同特征来确定查找的途径并选取所用的工

具。文献的特征有两个方面：外表特征和内容特征。所谓外表特征，是指文献的篇名（题名），著者姓名，文献序号（如科技报告号、技术标准号、专利号等），文种，发表年月，出版地点等；所谓内容特征，是指文献所研究的内容属于什么学科分支，新探讨的对象属什么主题，文献中提到了哪些关键词等。

（1）根据文献外表特征进行检索

A. 题名途径：根据以书刊名称或文章的篇名为特征编成的索引查找文献的途径。

B. 著者途径：根据著者的姓名来查找文献的途径，即查著者索引。

C. 文献类型的途径：文献类型有图书、期刊、科技报告、专利文献、政府出版物、技术标准、会议录等。文献检索工具根据需要也增设了文献类型的检索途径，就是报告号索引、专利号索引、会议录索引等。

（2）根据文献内容特征进行检索

A. 分类途径：按照文献主体内容所属的学科性质进行分类编排所形成的检索途径。

B. 主题途径：根据文献内容学科性质的主题进行检索的途径，即主题索引。

C. 关键词途径：关键词途径也称主题词途径，因为这种途径是通过关键词索引进行检索的。关键词索引又分为题内关键词索引、题外关键词索引和普通关键词索引。

D. 其他途径：根据不同的学科性质和特点以及不同的需要，编制独特的检索工具。

1·3 检索文献资料的方法

检索文献资料的方法有：追溯法、常用法、分段法。

（1）追溯法

所谓追溯法就是利用文章末尾所附的参考文献进行追溯查找的检索方法。这种查找方法就像滚雪球一样，随着"雪球"的滚动，获得的文献也越来越多。国外专门出版一种"引文索引"（Citation Index）。引文索引是查找引用论文的检索工具，是以某一特定的论文作者为检索途径，利用引文索引追溯查找。

Unit Twelve　Mechanical Engineering

（2）常用法

所谓常用法就是利用各种检索工具查找文献资料的方法。常用法又可分为顺查法、倒查法和抽查法。

A. 顺查法

顺查法是根据年代分析所得出的起始年代，利用选定的检索工具，在时间上由远及近地进行逐年查找的检索方法。

B. 倒查法

倒查法是利用选定的检索工具，在时间上由近及远地逐年逐卷地进行查找的检索方法。倒查法适用于新兴的研究课题，查找最近发表的文献。

C. 抽查法

抽查法就是在选定的课题研究发展兴旺的若干年中进行文献查找的检索方法。

（3）分段法

所谓分段法，也叫"循环法"或"混合法"，实际上是追溯法和常用法的综合。在查找文献资料时，既利用检索工具书刊查找文献资料，又利用文献资料后面所附的参考文献进行追溯，两种方法交替使用，直到查到所需资料为止。

在实际运用中，在检索工具比较齐全的情况下，主要采用常用法来查找文献资料。

1.4 检索文献资料的一般步骤

一般来说，利用检索工具查找文献资料的工作，可按下列步骤进行。

（1）分析研究所查找的课题，明确查找的目的与要求，确定所查找的文献的学科范围及类型。

（2）选择检索工具。

（3）确定检索途径和检索标志。

（4）利用检索工具查找并挑选文献。

（5）了解文献的馆藏情况并索取原始文献。

2 图书馆馆藏文献资料检索

通过上一节的介绍，我们了解了文献资料检索的基本知识，本节我们主要了解图书馆馆藏文献资料的检索。图书馆馆藏文献的主要类型有印刷型文献资料和非印刷型文献资料。下面简单介绍这两种文献资料的检索方法。

2.1 印刷型文献资料的检索

印刷型文献资料主要是指以纸为介质的载体所提供的资料，主要有中文图书、西文图书、中文期刊、西文期刊四大类。以南京航空航天大学图书馆为例，印刷型文献资料主要是通过"馆藏书刊目录查询"功能检索的。我们可以利用"馆藏书刊目录查询"功能查找图书，了解某一种图书是否被图书馆收藏，如果收藏，就可了解该书的馆藏地点、索书号和当前书刊状态。我们还可以利用"馆藏书刊目录查询"功能查找印刷版期刊。常用检索途径有：题名、ISSN号、文献号及关键词。书刊目录提供索取号（分类号）、条码号、年卷期、馆藏地、书刊当前状态等信息。

除了利用"馆藏书刊目录查询"功能进行检索，还可以利用各阅览室配备的书本式目录进行查询。如中文阅览室配备的社科期刊目录和科技期刊目录，它们分别按刊名汉语拼音的排列顺序排序和按中国图书分类法排序，以适应不同读者不同的检索习惯；外文阅览室的书本式目录是按照刊名字母顺序排列的。

2.2 非印刷型文献资料的检索

非印刷型文献资料主要是指以光、磁、电为存储手段存储在不同载体上的文献，如缩微型文献：缩微胶片、缩微胶卷、缩微卡片等；机读型文献：信息数据库资源、网络信息资源，以及磁带、磁盘、光盘等；视听型/声像型文献：电影片、幻灯片、录像带、唱片、录音带、CD、VCD等。

非印刷型文献资料一般可在图书馆非图书资料网络管理系统或电子阅览室、视听室查找到。

3 数据库文献资料检索

随着计算机信息技术的发展和文献资料的数字化，人们检索和利用信息的方式和途径也在发生变化，当今，数据库成为主要的计算机检索途径之一。因此，学会一些最基本的数据库检索技术有助于我们检索所需资料。

3·1 基本的检索技术

（1）布尔逻辑检索

布尔逻辑检索是多个检索项（可以是单词、词组或检索式）之间的布尔逻辑比较运算。分**逻辑与**、**逻辑或**、**逻辑非**三种。

逻辑与运算：逻辑与，算符 AND，也可写成 *，检索词 A 与检索词 B、C，用 AND 组配。检索词 A 与检索词 B、C 的提问表达式写作：A AND B AND C，或 A*B*C。

逻辑或运算：逻辑或，算符 OR，也可写成 +，检索词 A 与检索词 B、C，用 OR 组配。检索词 A 与检索词 B、C 的提问表达式写作：A OR B OR C，或 A+B+C。

逻辑非运算：逻辑非，算符 NOT，也可写成 –，检索词 A 与检索词 B、C，用 NOT 组配。检索词 A 与检索词 B、C 的提问表达式写作：A NOT B NOT C，或 A–B–C。

（2）限制检索

在检索系统中，通常有一些缩小检索结果的方法，即限制检索。常用的限制检索有两种方法：字段限制检索与使用限制符限制检索。

A. 字段检索

字段检索是一种限制检索词在某一字段范围出现的检索方法。数据库可供检索字段有两种：一种是反映内容的主题字段，如题名、叙词、标识词和文摘等；另一种是反映形式特征的检索字段，如作者、文献类型、语种、出版时间等。字段检索时，使用后缀符如："/ti""/de""/id""/ab"对主题字段加以限制，使用前缀符如："au=""py="等对非主题字段加以限制。

B. 使用限制符（+，-）限制检索

限制符限制检索是指检索结果中必须包含或者不包含某词语的场合，大多数系统都具有该项功能。把加号（+）放在一个词的前面表示在所有检索结果中都必须包含该词；把减号（-）放在一个词的前面表示在所有检索结果中都不能包含该词。

（3）截词检索

截词检索是一种在西文文献检索中常用的检索技术。截词检索方式可分为三种：后截断、前截断、中间截断。按截断的字符的数量，又可分有限截断与无限截断两种。通常用 * 表示无限截词。用 ? 表示有限截词。

后截断是最常用的一种检索技术，将截词符号放在一个字符串之后，以表示其后有限或无限个字符不影响之前的检索字符串的检索结果。如 **biolog***，可检索 biological、biologist、biology 等词；**physic?**，可检索 **physics**、**physical**、**physicist** 等词。

前截断将截词符号置于一个字符串的前方以表示其之前有限或无限个字符不影响之后的检索字符串的检索结果，如 ***physics**，可检索 **physics**、**astrophysics**、**biophysics**、**geophysics** 等词。

中间截断也叫"内嵌字符截断"。将截词符号置于一个检索词中间，而不影响前后字符串的检索结果。如 **organi?ation**，可检索 **organization**、**organisation**。

（4）原文检索

原文检索是以原始记录的检索词与检索词间特定位置关系为对象的运算。它是一种不依赖叙词表而直接进行自由词检索的一种方法。它的运算类型主要有如下四种：

A. 记录级检索，要求检索词出现在同一记录中。

B. 字段级检索，要求检索词出现在同一字段中。

C. 子字段或自然句级中检索，要求检索词出现在同一子字段或同一自然句级中。

D. 词位置检索，要求检索词之间的相互位置满足某些条件。

（5）模糊检索

模糊检索允许被检索信息和检索提问之间存在一定的差异，这种差异就

Unit Twelve　Mechanical Engineering

是"模糊"在检索中的含义。比如，想查找"计算机维修"的信息，不知其在数据库中的标引词是什么，输入"计算机的维修"或"计算机维修"都能得到想要的结果。模糊检索中所指的差异一方面来自用户在检索提问时的输入错误；另一方面来自某些词在不同国家的不同形式。

3·2 不同的数据库的使用

不同的数据库包含的内容不完全一样，所提供的检索途径也有所不同，但无论什么类型的文献信息数据库，其基本特征是相同的。数据库种类很多，如南京航空航天大学图书馆部分数据库（表1），我们可以尝试在不同的数据库中查找所需的文献资料。

表 1　南京航空航天大学图书馆部分数据库

	电子图书	电子期刊	学位论文	会议论文	综合数据库
中文	•超星数字图书馆 •书生之家 •南航教学参考书资源库（全文）	•中国学术期刊网 •维普中文科技期刊 •万方数字化期刊	•万方中国学位论文数据库（全文） •南航博硕士学位论文（全文）	•万方中国会议论文数据库（全文）	•万方科技信息 •中国引文数据库 •中国人大复印资料全文数据库 •中文社会科学引文索引（文摘）
西文	•Springer全文电子图书 •北京金图国际外文电子图书（全文）	•EBSCO、Elsevier、Kluwer、Springer、John Wiley综合类学术期刊（全文） •Science Online 科学在线	•PQDD博硕士论文文摘库 •PQDD博硕士论文全文库		•高校英语资源总库 •FirstSearch联机检索 •DIALOG数据库 •专业数据库 •文摘数据库

4　网络文献资料检索

互联网上有很多搜索文献的方法，但怎样查找才能更准确、更迅速呢？一种有效的方式是使用搜索引擎（Search Engine）。

搜索引擎其实是互联网上的一类网站，它事先将网上各个网站的信息进行分类并且建立索引，然后把内容索引存放到一个地址数据库中，当人们向搜索引擎发出搜索请求时，搜索引擎便在其数据库中搜索，找到一系列相关信息后，将结果以网页的形式返回。

搜索引擎按其工作方式主要可分为两种。一种是全文搜索引擎，它是名副其实的搜索引擎，国外具代表性的有 AltaVista、Google、Excite、Hotbot、Lycos、Fast/AllTheWeb、Overture，国内常用的引擎有百度。它们都是通过在互联网上提取各个网站的信息来建立自己的数据库，并向用户提供查询服务的，因此是真正的搜索引擎。另一种是目录索引（Search Index/Directory）。实际上它们算不上是搜索引擎，仅仅是按目录分类的网站链接列表而已。用户完全可以不用进行关键词查询，仅靠分类目录也可找到需要的信息。目录索引中具有代表性的有 Yahoo、LookSmart、About、Ask Jeeves，国内的搜狐、新浪、网易搜索也都属于这一类。

从浩瀚的互联网快捷准确地提取需要的信息，需要一定的策略和技巧。下面简单介绍一些搜索策略和技巧。

4.1 选择合适的搜索工具

不同目的的查询应当选用不同的搜索引擎。每种搜索引擎都有各自的特点，只有选择合适的搜索工具才能得到最佳的结果。那么，全文搜索引擎和目录索引这两种搜索工具哪种更好呢？这取决于你想查询的问题。一般而言，如果你查找非常具体或者特殊的问题，就需要用全文搜索引擎，如 Google；如果你希望浏览某方面的信息或者专题，那么类似 Yahoo 的分类目录可能会更合适；如果你需要查找的是某些确定的信息，最好使用专门的搜索引擎。下面介绍几个常用的专门搜索引擎：Google Scholar (http://scholar.google.com/)，谷歌学术，用于搜索学术文献，包括同行评议的论文、学位论文、图书、预印本、技术报告等，涉及各学科领域；Google Book Search (http://books.google.com/)，谷歌的书籍搜索服务；Scirus (http://www.elsevier.com/)，是由爱思唯尔科学公司推出的科学专用搜索引擎，是目前网上最全面、综合性最强的文献搜索引擎之一；虚拟图书馆 (http://vlib.org/)，是以主题树或数据库结合超文本链接的方式供用户浏览或者查询的搜索引擎；Scicentral (http://

www.scicentral.com/），是一个非常好的综合专业搜索引擎，它搜集了大量的各类期刊、追踪各学科最新的研究报告，并为用户提供网上免费服务。

4·2 在搜索引擎中运用查询技巧

在使用搜索引擎查找信息的过程中，经常会遇到由于返回的结果太多，而很难发现有用信息的情况，因此掌握常用搜索引擎的使用技巧非常必要。多数搜索引擎都有介绍该搜索工具使用技巧的信息，以便使用者能够快速掌握基本技巧。要全面了解有关搜索引擎的知识，我们可以到中文搜索引擎指南（http://www.sowang.com/）以及中国搜索（http://www.chinaso.com/）等网站学习。了解了搜索引擎的相关知识，自然会产生事半功倍的效果。

搜索引擎返回的结果通常是相关信息的链接，很多时候由于某种原因这些链接不能访问，这时可以试试搜索引擎提供的"网页快照"或"历史网页"这一类功能。

4·3 运用语法规则进行组合搜索

搜索引擎支持许多附加逻辑命令查询，不同的搜索引擎的逻辑命令也不完全相同。常用的搜索引擎语法及命令有"＋"（与）、"OR"（或）和"－"（非），正好对应布尔（Boolean）逻辑命令 AND（与）、OR（或）和 NOT（非）。用好这些命令符号可以大幅提高我们的搜索精度。

通常，使用单个关键词搜索到的信息条目数量庞大，其中大部分并不符合我们的要求。假如在搜索时使用两个或多个关键词，在多个关键词之间输入"＋"，表示逻辑"与"操作，就可以提高搜索结果的精确度。例如，搜索与"电器"和"维修"有关的信息，只需输入"电器＋维修"即可。

在多个关键词之间输入"－"，表示逻辑"非"操作。"A-B"表示搜索包含 A 但不包含 B 的网页。例如，输入"电器＋维修－汽车"，表示搜索所有包含"电器"和"维修"，但不包含"汽车"的信息。

"OR"表示逻辑"或"操作。"A OR B"表示在搜索的网页中，或有 A，或有 B，或兼有 A 和 B。例如，输入"电器 OR 维修"，表示搜索标题中含有"电器"或"维修"的网页。

4.4 使用特殊搜索命令

在搜索文献过程中，有时难免会有一些特殊的需求，而搜索引擎也支持一些特殊的搜索命令，以方便我们精确定位所需信息。

（1）标题搜索

"intitle:"是多数搜索引擎都支持的针对网页标题的搜索命令，前文提到的逻辑命令同样适用于标题搜索。例如，输入"intitle：电器"，表示要搜索标题含有"电器"的网页。返回的结果都是标题中包含关键字、词的信息条目。

（2）网站搜索

我们还可以针对网站进行搜索。"site:"表示搜索结果局限于某个具体网站或网站频道。例如，输入"电器＋维修 site:com.cn"，表示在域名中含有"com.cn"的网站中搜索含有"电器"和"维修"的信息。

（3）链接搜索

在 Google 和 AltaVista 中，用户均可通过"link:"命令来查找某网站的外部导入链接。

（4）其他搜索命令

除上述命令外，还有其他一些特殊搜索命令，如"filetype:"（限定搜索的文档类别）、"daterange:"（限定搜索的时间范围）、"phonebook:"（查询电话）等。

搜索引擎的搜索技术在飞速发展，搜索技巧也在不断变化，我们需要不断地积累经验，让搜索引擎成为我们检索文献的好帮手。

Unit Twelve   Mechanical Engineering

Quiz

1. **Use the search engine to get information.**

    (1) Enter the Internet and see how many sites you can get from the search engine by just typing the word "science". Then search for other subjects on science, for example, science history, science research, etc. What does the result tell you about science?

    (2) Enter the Internet and see how many sites you can get from the search engine by just typing the word "life science" "earth science" "physical science". What does the result tell you about?

    (3) Collect a magazine article or news clipping on an important question that scientists are currently investigating. Try to find out more about current science research on the Internet.

Part B Reading

The Engineering Profession

Engineering is one of the oldest occupations in history. Without the skills included in the broad field of engineering, our present-day civilization never could have evolved.[1] The first tool-makers who chipped arrows and spears from rock were the **forerunners** of modern mechanical engineers. The **craftsmen** who discovered metals in the earth and found ways to refine and use them were the ancestors of mining and **metallurgical** engineers. And the skilled technicians who devised irrigation systems and erected the marvelous buildings of the ancient world were the civil engineers of their time. One of the earliest names that has come down to us in history is that of Imhotep, the designer of the stepped pyramid at Sakkara in Egypt about 3000 BC.

Engineering is often defined as making practical application of theoretical sciences such as physics and mechanics. Many of the early branches of engineering were based not on science but on **empirical** information that depended on observation and experience rather than on theoretical knowledge. Those who devised methods for splitting the massive blocks that were needed to build Stonehenge in England or the unique pyramids of Egypt discovered the principle of the **wedge** by trial for the pyramids were probably raised into place by means of **ramps** of earth that surrounded the structure as they rose; it was a practical application of the **inclined plane**, even though the concept was not understood in terms that could be quantified or expressed mathematically.[2]

Quantification has been one of the principal reasons for the explosion of scientific knowledge since the beginning of the modern age in the sixteenth and seventeenth centuries. Another important factor has been the development of the experimental method to **verify** theories.

Unit Twelve Mechanical Engineering

Quantification involves putting the data or pieces of information resulting from experimentation into exact mathematical terms. It cannot be stressed too strongly that mathematics is the language of modern engineering.[3]

Since the nineteenth century both scientific and practical applications of its results have **escalated**. The mechanical engineer now has the mathematical ability to calculate the mechanical advantage that results from the complex interaction of many different mechanisms. He or she has new and stronger materials to work with and enormous new sources of power. The Industrial Revolution began by putting water and steam to work; since then machines using electricity, gasoline, and other energy sources have become so widespread that they now do a very large proportion of the work of the world.

One result of the rapid expansion of scientific knowledge was an increase in the number of scientific and engineering specialties. By the end of the nineteenth century not only were mechanical, civil, and mining and metallurgical engineering established but the newer specialties of chemical and electrical engineering emerged. This expansion has continued to the present day. We now have, for example, nuclear, petroleum, aerospace, and electronic engineering. Within the field of each engineering there are subdivisions. For example, within the field of civil engineering itself, there are subdivisions: structural engineering, which deals with permanent structure; hydraulic engineering which is concerned with systems involving the flow and control of water or other fluids; and **sanitary** or environmental engineering, which involves the study of water supply, purification, and **sewer** systems. The major subdivision of mechanical engineering is industrial engineering which is concerned with complete mechanical systems for industry rather than individual machines.

Another result of the increase in scientific knowledge is that engineering has grown into a profession. A profession is an occupation like law, medicine, or engineering that requires specialized, advanced education; indeed, they are often called the "learned professions". Until the nineteenth century, engineers generally were craftsmen or project organizers who learned their skills through **apprenticeship**, on-the-job training, or trial and error. Nowadays, many engineers spend years studying at universities for advanced degrees. Yet even those engineers who do not study for advanced degrees must be aware of changes in their field and those related to it.

Thus, the word engineer is used in two ways in English. One usage refers to the professional engineer who has a university degree and education in mathematics, science, and one of the engineering specialties. Engineer, however, is also used to refer to a person who operates or maintains an engine or machine. An excellent example is the railroad locomotive engineer who operates a train. Engineers in this sense are essentially technicians rather than professional engineers.

New Words and Expressions

forerunner 先驱（者），先锋
craftsmen 工匠，技工
metallurgical 冶金的
empirical 经验主义的，以经验为根据的
wedge 楔
ramp 斜坡
inclined plane 斜面
verify 检验
escalate 逐步上升
sanitary （环境）卫生的
sewer 污水管，下水道
apprenticeship 学徒

Notes

1. Without..., our present-day civilization never could have evolved. 句中虚拟语气表示本来应该发生但却没有发生的事情。

2. Those who...that could be quantified or expressed mathematically. 本句为并列复合句，宜采用断句法翻译。本句译为：那些当时想出办法把建造英国巨石阵和世界独有的埃及金字塔所需要的巨石劈开的

Unit Twelve　Mechanical Engineering

人，在试验中发现了楔的原理，因为金字塔很可能是在建筑的过程中利用环绕在四周的斜土坡逐步堆砌上去的；这是斜面原理的一个实际应用，尽管人们在当时对斜面原理的概念还不能从量上去了解，也不能用数学公式去表示。

3. cannot be stressed too strongly 意为"怎么（做）都不会太……""无论怎样（做）都不算太……"。

Exercises

1. Find out the English equivalents of the following engineering terms from the passage.

   （1）机械工程　　　　　　（2）土木工程
   （3）采矿工程　　　　　　（4）冶金工程
   （5）化学工程　　　　　　（6）电气工程
   （7）结构工程　　　　　　（8）水力工程
   （9）（环境）卫生工程　　（10）环保工程
   （11）工业工程　　　　　（12）核工程
   （13）石油工程　　　　　（14）航空航天工程
   （15）电子工程

2. Find out the English equivalents of the following Chinese terms from the passage.

   （1）理论知识　　　　　　（2）量化
   （3）科学知识爆炸　　　　（4）实验方法
   （5）供水　　　　　　　　（6）净化
   （7）排污系统　　　　　　（8）工程项目的组织者
   （9）专业　　　　　　　　（10）岗位培训

3. Translate the following sentences into Chinese.

   (1) Engineering is one of the oldest occupations in history. Without the skills included in the broad field of engineering, our present-day

civilization never could have evolved.

(2) Engineering is often defined as making practical application of theoretical sciences such as physics and mechanics. Many of the early branches of engineering were based not on science but on empirical information that depended on observation and experience rather than on theoretical knowledge.

(3) It cannot be stressed too strongly that mathematics is the language of modern engineering.

(4) The mechanical engineer now has the mathematical ability to calculate the mechanical advantage that results from the complex interaction of many different mechanisms.

(5) A profession is an occupation like law, medicine, or engineering that requires specialized, advanced education; indeed, they are often called the "learned professions".

Part C Extended Reading

Functions of Mechanical Engineering

Four functions of the mechanical engineering are cited. The first is the understanding of and dealing with the bases of mechanical science. These include dynamics, concerning the relation between forces and motion, such as in **vibration**; automatic control; thermodynamics, dealing with the relations among the various forms of heat, energy, and power; fluid flow; heat transfer; lubrication; and properties of materials.

Second is the **sequence** of research, design, and development. This function attempts to bring about the changes necessary to meet present and future needs. Such work requires not only a clear understanding of mechanical science and an ability to analyze a complex system into its basic factors, but also the originality to synthesize and invent.

Third is production of products and power, which **embraces** planning, operation, and maintenance. The goal is to produce the maximum value with the minimum investment and cost while maintaining or enhancing longer-term **viability** and reputation of the **enterprise** or the institution.

Fourth is the coordinating function of the mechanical engineering, including management, consulting, and, in some cases, marketing.

In all of these functions there is a long continuing trend toward the use of scientific instead of traditional or intuitive methods, an aspect of the ever-growing professionalism of mechanical engineering. Operations research, value engineering, and PABLA (problem analysis by logical approach) are typical titles of such new rationalized approaches. Creativity, however, cannot be rationalized. The ability to take the important and unexpected step that opens up new solutions remains in mechanical engineering, as elsewhere, largely a personal and spontaneous characteristic.

New Words and Expressions

vibration 振动
sequence 次序，顺序
embrace 包括，包含
viability 生存能力
enterprise 企业

Exercises

1. Find out the English equivalents of the following Chinese terms from the passage.

   （1）动力学　　　　　　　（2）自动控制
   （3）热力学　　　　　　　（4）流体流动
   （5）热传递　　　　　　　（6）润滑
   （7）材料特性　　　　　　（8）开发
   （9）运作　　　　　　　　（10）维护
   （11）最大的价值　　　　　（12）最少的投资
   （13）管理　　　　　　　　（14）咨询
   （15）市场营销　　　　　　（16）运筹学
   （17）工程经济学　　　　　（18）逻辑法问题分析

2. Translate the following sentences into Chinese.

   (1) The first is the understanding of and dealing with the bases of mechanical science.

   (2) Such work requires not only a clear understanding of mechanical science and an ability to analyze a complex system into its basic factors, but also the originality to synthesize and invent.

   (3) The goal is to produce the maximum value with the minimum investment and cost while maintaining or enhancing longer-term viability and reputation of the enterprise or the institution.

   (4) In all of these functions there is a long continuing trend toward the use of scientific instead of traditional or intuitive methods, an aspect

of the ever-growing professionalism of mechanical engineering.

(5) The ability to take the important and unexpected step that opens up new solutions remains in mechanical engineering, as elsewhere, largely a personal and spontaneous characteristic.

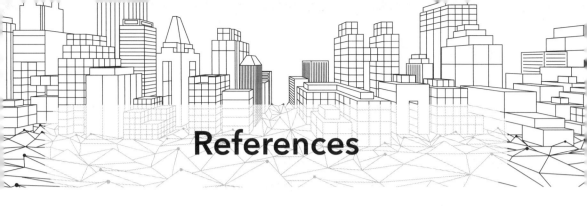

References

艾秋. 2002. 英语听力好易通：慢速英语. 北京：世界知识出版社.

卜玉坤. 2000. 大学专业英语：计算机英语（第一册）. 北京：外语教学与研究出版社.

卜玉坤. 2001. 大学专业英语：机械英语（第一册）. 北京：外语教学与研究出版社.

蔡忠元，吴鼎民. 2000. 航空专业英语教程. 北京：航空工业出版社.

陈秋劲，Richard B. Baldauf, David Gordon Etheridge. 2005. 英汉互译理论与实践. 武汉：武汉大学出版社.

陈晓莹. 2001. 新世纪英语阅读套餐：知识篇. 重庆：重庆大学出版社.

陈新. 1999. 英汉文体翻译教程. 北京：北京大学出版社.

陈羽伦. 1995. 科技英语选粹. 北京：中国对外翻译出版公司.

程秀芳，周丽娟. 2006. 身边的科技英汉对照读物. 北京：中国水利水电出版社.

戴丹青，胡晓军. 2001. 读点科学史. 上海：上海科技教育出版社.

戴炜华，陈文雄. 1984. 科技英语的特点和应用. 上海：上海外语教育出版社.

丁树德. 2005. 翻译技法详论. 天津：天津大学出版社.

范武邱. 2001. 实用科技英语翻译讲评. 北京：外文出版社.

方梦之，毛忠明. 2005. 英汉—汉英应用翻译教程. 上海：上海外语教育出版社.

冯志杰. 1998. 汉英科技翻译指要. 北京：中国对外翻译出版公司.

高凤平. 2006. CET-6 科普阅读. 北京：世界图书出版公司.

郭富强. 2004. 英汉翻译理论与实践. 北京：机械工业出版社.

郭建中. 2004. 科普与科幻翻译：理论、技巧与实践. 北京：中国对外翻译出版公司.

何继善. 2000. 现代科技写作. 长沙：湖南人民出版社.

李兆平，仲锡. 2002. 说科学. 北京：华文出版社.

林钜洸. 2004. 当代高科技及其哲理. 北京：外语教学与研究出版社.

刘宓庆. 1998. 文体与翻译. 北京：中国对外翻译出版公司.

马锦儒，范淑芹，张现彬. 2003. 高级百科读物英汉对照 50 篇. 上海：上海外语教育出版社.

马锦儒，张现彬. 2001. 中级百科读物英汉对照 100 篇. 上海：上海外语教育出版社.

彭宣维. 2000. 英汉语篇综合对比. 上海：上海外语教育出版社.

申雨田，戴宁. 2002. 实用英汉翻译教程. 北京：外语教学与研究出版社.

司树森，刘典忠. 2005. Super·分级分类英汉对照读物·四级：科普篇. 北京：中国人民大学出版社.

索恩利. 1981. 英语科普文选（第三集）. 京广英，译. 北京：科学普及出版社.

谭卫国. 2000. 英语背诵范文精华. 上海：华东理工大学出版社.

田传茂，许明武，杨宏. 2003. 科技英语阅读系列：生命科学. 武汉：华中科技大学出版社.

王建武，李民权，曾小珊. 2000. 科技英语写作——写作技巧·范文. 西安：西北工业大学出版社.

王令坤，朱俊松，朱慧敏，李振国，葛纪红. 2002. 科技英语读写译教程. 北京：外语教学与研究出版社.

王汝琳. 2004. 科技英语阅读精选（计算机，电子信息，航空航天）. 北京：化学工业出版社.

王守仁. 2016.《大学英语教学指南》要点解读. 外语界（3）：2–10.

王秀芝，王秀盈. 2002. 生物—医学英汉时文精粹. 北京：世界知识出版社.

王佐良，丁往道. 1987. 英语文体学引论. 北京：外语教学与研究出版社.

魏巍，周成. 2003. 航空航天技术. 北京：国防工业出版社.

翁凤翔. 2002. 实用翻译. 杭州：浙江大学出版社.

吴寒，刘子毅. 2003. 实用英语应用写作. 广州：中山大学出版社.

谢小苑. 2008. 科技英语翻译技巧与实践. 北京：国防工业出版社.

References

谢小苑. 2015. 科技英语翻译. 北京：国防工业出版社.

徐世延. 1985. 基础科技英语教程. 上海：上海科学技术大学.

严俊仁. 2000. 科技英语翻译技巧. 北京：国防工业出版社.

《英语学习》编辑部. 2002. 科技之虹. 北京：外语教学与研究出版社.

俞宝发. 2002. 英语背诵范文精典. 上海：生活·读书·新知三联书店.

袁亦宁. 2004. 民航英语核心读本. 北京：北京航空航天大学出版社.

约翰·斯伟尔斯. 1980. 科技英语写作法. 颜绍知，译. 北京：原子能出版社.

郑福裕. 2003. 科技论文英文摘要编写指南. 北京：清华大学出版社.

钟似璇. 2004. 英语科技论文写作与发表. 天津：天津大学出版社.

Boden. M. A. 2018. *Artificial Intelligence: A Very Short Introduction.* Oxford: Oxford University Press.

Doo Hwan Kim. 2008. *Essays for the Study of the International Air and Space Law.* Seoul: Korean Studies Information.

Ferguson. R. 2007. Civil aviation faces green challenge. *Nature* (448): 120-121.

O'Reilly Team. 2011. *Big Data Now: Current Perspectives from O'Reilly Radar.* California: O'Reilly Media, Inc.

Key and Reference for Translation

Unit One　Science and Technology

Part A　Lecture

Quiz

1. （1）用滚动摩擦代替滑动摩擦，会大幅度减小摩擦力。
   （2）天然橡胶取自橡胶树，是一种叫"乳胶"的白色乳状液体。在运往世界各国之前，橡胶要经过酸处理和烘干。
   （3）今天，电子计算机广泛地用于解决一些数学问题，这些问题与天气预报和把卫星送入轨道有关。
   （4）当蒸汽重新冷却为水时，所释放出的热量与其原来形成蒸汽时所吸收的热量相等。
   （5）热能在辐射时转换成性质与光相似的辐射能。
   （6）由于哈勃发明了利用空气中的氮气的方法，这种局面就完全改观了。
   （7）普通的现代人看物体时不需要这种显微镜般的精密程度。
   （8）这是一种电学方法，它最有希望用于含盐分的水。
   （9）以前人们认为原子是最小的物质单位，现在才知道原子还可以分为原子核与电子、中子与质子等。
   （10）如果冷却系统中没有空气存在，冷却效果就不会受到影响，温度也不可能保持这样低。
   （11）实验表明，使电流流动的电压，电流流动的速率与电流所通过的物质的电阻这三者之间确有一定的相互关系。
   （12）为了解释光学现象，人们曾试图假定有一种具有与弹性固体相同的物理性的介质。这种尝试的结果，最初曾使人们了解到一种能传输横向振动的介质的具体实例，但后来却使人得出了这样一个明确的结论：并不存在任何具有上述假定所认为的那种物理性质的发光介质。

Part B　Reading

Exercises

1. (1) pure science　　　　　　　　(2) applied science
   (3) model　　　　　　　　　　　(4) hypothesis
   (5) the working laws of science　　(6) the life cycle
   (7) equation　　　　　　　　　　(8) radioactivity
   (9) metal-fatigue　　　　　　　　(10) anthropology

2. （1）理论科学主要关注的是那些建立宇宙现象之间联系的理论的发展（或者通常称为模型）。

   （2）在从事这方面工作时，理论科学家通常并不顾及（考虑）理论的实际应用问题，而只是力图去解释事情发生的方式和原因。

   （3）有关材料强度和应用的研究，将理论数学上的发现应用并推广到农业和社会科学的取样过程中，原子能潜能的开发活动中等，都是应用科学家或技术人员从事科学应用方面工作的例证。

   （4）显然，应用科学的许多分支实际上都是纯理论研究或实验研究的延续。

   （5）由此看来，科学上的这两大范畴似乎是相互依存、彼此促进的，而理论科学家与应用科学家之间的所谓区分，在实质上并不那么泾渭分明。

Part C　Extended Reading

Exercises

1. (1) physical sciences　　　　　　(2) earth sciences

(3) life sciences
(4) astronomy
(5) paleontology
(6) oceanography
(7) meteorology
(8) botany
(9) zoology
(10) genetics
(11) mechanics
(12) thermodynamics
(13) astrophysics
(14) horticulture
(15) animal husbandry

2.（1）的确，数学常常被称为科学的语言。它是传达科学结果最重要、最客观的手段。

（2）除了这些独立的学科，还有许多由两个或多个分支组成的学科领域，如天文物理学、生物物理学、生物化学、地质化学和地质物理学等。

（3）所有这些研究领域都可称为理论科学，与它相对应的是应用科学或工程科学，也就是技术。应用科学关注科学活动成果的实际应用情况。

参考译文

理论科学与应用科学（Pure and Applied Science）

理工科学生有时候也许会被"理论"科学与"应用"科学这两个名称弄得莫名其妙。难道这两者真的像字面所表明的那样，有着截然不同的内容，很少相关，甚至完全不相干吗？让我们先来看一看两者各自的作用吧。

理论科学主要关注的是那些建立宇宙现象之间联系的理论的发展（或者通常称为模型）。一旦有效，这些理论（假设、模型）便成为现行的科学定律或原理。在从事这方面工作时，理论科学家通常并不顾及（考虑）理论的实际应用问题，而只是力图去解释事情发生的方式和原因。所以，物理学上描述基本粒子运转状态的方程式，或生物学上研究那些生活在极地严寒环境里的特殊昆虫的生命周期，都可以说是理论科学（基础研究）的实例。在那种情况下，这些研究与技术，与应用科学并无明显的联系。

另一方面，应用科学却直接关联到把理论科学现行的定律应用于实际生活之中，以增强人类对环境的控制能力，从而导致新技术的发展、新工艺的制作和新机器的研制。有关材料强度和应用的研究，将理论数学上的发现应用并推广到农业和社会科学的取样过程中，原子能潜能的开发活动中等，都是应用科学家或技术人员从事科学应用方面工作的例证。

显然，应用科学的许多学科分支实际上都是纯理论研究或实验研究的延续。所以，放射性的研究始于理论研究，但其成果在今天却以种种不同的方式得到应用：医学上用于治疗癌症，农业上用于研制肥料，工程学上用于研究金属疲劳，人类学和地质学等方面，则把它应用在估算研究对象的寿龄或年代的方法中。相反地，应用科学和技术工作又常常对理论科学的发展起着直接的推动作用。比方说，在把理论科学某一特定概念应用于某实际问题时，技术人员揭示了该理论模型的缺陷或局限性，从而为进一步的基础研究指明了方向，这时候，进一步的相互作用就显现出来了，因为理论科学家还要依靠别的技术人员给他提供更先进的仪器，以进行进一步的基础研究。

由此看来，科学上的这两大范畴是相互依存、彼此促进的，而理论科学家和应用科学家之间的所谓区分，在实质上并不那么泾渭分明。

科学的分支（Branches of Science）

科学大致可以被划分为物理科学、地球科学和生命科学。虽然数学本身不是一门科学，却由于被科学广泛使用而与其产生密切关联。的确，数学常常被称为科学的语言。它是传达科学结果最重要、最客观的手段。物理科学包括物理、化学和天文学；地球科学（有时被看作物理科学的一部分）包括地质学、古生物学、海洋学和气象学；生命科学包含所有生物学的分支，如植物学、动物学、遗传学和医学。每一个学科本身又可以再分成不同的分支，如数学可以分成算术、代数、几何与数学分析；物理可以分为力学、热力学、光学、声学、电学和磁学，以及原子物理学和核物理学。除了这些独立的学科，还有许多由两个或多个分支组成的学科领域，如天文物理学、生物物理学、生物化学、地质化学和地质物理学等。

所有这些研究领域都可称为理论科学，与它相对应的是应用科学或工程科学，也就是技术。应用科学关注科学活动成果的实际应用情况。应用科学领域包括机械工程、土木工程、航空工程、电力工程、建筑工程、化学工程及其他工程；也包括农学、园艺学和畜牧学，以及医学的诸多方面；最后，还有研究科学历史和哲学等的独特学科。

Unit Two　Science and Technology in the Past

Part A　Lecture

Quiz

1. （1）科学的发现与发明，并不总是会对语言产生与其重要性相当的影响。
   （2）超导体在电器应用上的重要性不容小觑。
   （3）因此，我们说，除了电压外，还有一些因素决定电流的大小。
   （4）如果早些采用遥测技术，许多有发展前景的飞机研究工作原本是不会被取消的，某些飞机坠毁的事故也可以避免。
   （5）指示灯熄灭，表示操作终止。
   （6）如果设计者把应当设计为 1,000 磅的托架设计成 100 磅，那么肯定会出现事故。
   （7）只要有足够的电位差，电流便可通过任何物体。
   （8）炉壁周围采用耐火砖可大大降低热耗。
   （9）电视通过无线电波发射和接收各种活动物体的图像。
   （10）后来，他发现在太阳西部的边缘上有一群黑子，过了好一会儿，黑子渐渐消失，出现在太阳的东部边缘，最后又恢复原位。

Part B　Reading

Exercises

1. (1) printing　　　　　　　　　(2) gunpowder
   (3) magnetism　　　　　　　　(4) structure of Chinese society
   (5) block printing　　　　　　 (6) Buddhist text
   (7) classical books of Confucius　(8) moveable type
   (9) magnetic compass　　　　　(10) south-controlling spoon
   (11) magnetic declination　　　 (12) magnetic polarity

2.（1） 一直以来，西方人都普遍认为，古代中国的科学技术同欧洲相比是微不足道的。

（2） 印刷术、火药和指南针这三大发明在中国的出现都要早于欧洲，但与欧洲不同，这些发明创造并未引起中国社会结构的重大变革。

（3） 活字印刷出现于11世纪，虽然当时成千上万的汉字必须各有一块印模。直到400年后的1456年，谷登堡用活字印刷术印刷了其拉丁文《圣经》时，活字印刷才在欧洲出现。

（4） "指南勺"由一块天然磁石雕刻而成。当置于高度抛光的铜板上时，它会一直转动，直到指向南方。

（5） 看来这些磁力罗盘很有可能早在10世纪就已用于航海，而且有证据表明中国人在欧洲人知道磁的二极性之前就知道了磁偏角，即罗盘并不正好指向南北，并且其角度差异随时间而变化。

Part C　Extended Reading

Exercises

1.（1） 在宋朝时期（960年—1279年），中国复杂的经济制度促使纸币等发明的诞生。10世纪火药的发明推动了一系列发明的产生，如火枪、地雷、水雷、手炮、炮弹、多节火箭以及带有空气动力机翼和爆炸物的火箭弹等。

（2） （英国汉学家）白馥兰说，龙山文化时期（约前3000年—约前2000年），人们驯养牛和水牛，当时还没有灌溉术，也没有高产作物，且有充足的证据表明，当时人们已经种植旱地谷物，而旱地谷物"只有在精心耕种下"才能获得高产，这揭示了中国早在龙山文化时期就懂得使用犁，这也解释了农业高产的原因，使得中华文明在商朝时期（约前1600年—前1050年）兴起。

（3） 本文所列均为首先在中国发明的技术，所以并不包括中国与外国接触过程中传入的技术，如来自中东的风车或现代欧洲早期的望远镜。

Key and Reference for Translation

参考译文

古代中国的科学和技术（Science and Technology in Traditional China）

一直以来，西方人都普遍认为，古代中国的科学技术同欧洲相比是微不足道的。现在我们知道事实并非如此。古代中国已经形成了一个涉及众多科技领域的庞大知识体系，其中的许多知识都先于欧洲，有的甚至比欧洲早了好几个世纪；而且，这些知识均产生于一个对欧洲的一切知之甚少的社会之中。

印刷术、火药和指南针这三大发明在中国的出现都要早于欧洲，但与欧洲不同，这些发明创造并未引起中国社会结构的重大变革。

刻版印刷在中国出现于公元 9 世纪，刻版印刷的书籍则出现于 9 世纪后期。现存最早的刻版印刷书籍是一部佛经，时间为公元 868 年；一套完整的儒家经典著作也于公元 932 年交付印刷，并于公元 953 年出版。活字印刷出现于 11 世纪，虽然当时成千上万的汉字必须各有一块印模。直到 400 年后的 1456 年，谷登堡用活字印刷术印刷了其拉丁文《圣经》时，活字印刷才在欧洲出现。

火药在中国也是起源于公元 9 世纪。在一本出版于 1044 年的中文书籍中最早以书面形式记载了一种用木炭、硝石和硫黄混合制成火药的方法。直到 14 世纪初，欧洲才有类似的记载。然而，这一新发明很快就被用于火箭发射器和火药枪等武器。

首次记载磁力和类似磁力罗盘的时间甚至更早。在一本可追溯到公元 83 年出版的文献中，曾经提到过一个"指南勺"。它由一块天然磁石雕刻而成。当置于高度抛光的铜板上时，它会一直转动，直到指向南方。中国的罗盘总是指向南方！在后来的几个世纪中，出现了许多有关"指南针"的记载——远远早于欧洲在 1180 年首次提及磁的二极性。

看来这些磁力罗盘很有可能早在 10 世纪就已用于航海，而且有证据表明在欧洲人知道磁的二极性之前中国人就知道了磁偏角，即罗盘并不正好指向南北，并且其角度差异随时间而变化。

中国的发明（List of Chinese Inventions）

中国是许多重大发明的发源地，包括中国古代的四大发明：造纸术、指南针、火药和印刷术（包括雕版印刷和活字印刷）。以下列表包含上述及其他发明。

中国人发明了涉及机械学、水利学和数学的技术，应用于计时、冶金、天文、农业、工程、乐理、技艺、航海和战争。时至战国时期（前403年—221年），中国已拥有先进的冶金技术，包括高炉和化铁炉；而汉代（前206年—220年）已掌握了百炼钢与炒钢技术。在宋朝时期（960年—1279年），中国复杂的经济制度促使纸币等发明的诞生。10世纪火药的发明推动了一系列发明的产生，如火枪、地雷、水雷、手炮、炮弹、多节火箭以及带有空气动力机翼和爆炸物的火箭弹等。随着11世纪指南针在航海上的应用，以及使用1世纪发明的船尾舵在公海上航行的能力，近代中国海员们远航至东非和埃及。至于水力钟表结构，中国人自8世纪就已使用擒纵机械，11世纪使用环状传动链条。中国人还建造了由水车舵轮驱动的大型机械木偶剧院，以及由桨轮驱动的侍酒机器人。

同时期的裴李岗文化和彭头山文化形成于公元前7000年，是中国最古老的新石器文化的象征。史前中国新石器时期的一些最早发明包括半月形和长方形石刀、石锄和石锹、小米、大米和大豆种植，养蚕缫丝，石灰地板的夯土结构建筑，陶工旋盘、篮子状陶器、三脚陶器和陶制蒸锅，以及礼器和肩胛骨占卜术。（英国汉学家）白馥兰说，龙山文化时期（约前3000年—前2000年），人们驯养牛和水牛，当时还没有灌溉术，也没有高产作物，且有充足的证据表明，当时人们已经种植旱地谷物，而旱地谷物"只有在精心耕种下"才能获得高产，这揭示了中国早在龙山文化时期就懂得使用犁，这也解释了农业高产的原因，使得中华文明在商朝时期（约前1600年—前1046年）兴起。后来多管式条播机和厚板铁犁的发明，使得中国农作物产量足以养活更多的人口。

本文所列均为首先在中国发明的技术，所以并不包括中国与外国接触过程中传入的技术，如来自中东的风车或现代欧洲早期的望远镜。本文所列亦不包括其他地方开发、后由中国单独发明的技术，如里程表和链式泵。科学、数学或自然发现，设计或风格的细微改变以及艺术创新等不能被视为发明，所以本文也不赘述。

Unit Three Language of Science

Part A Lecture

Quiz

1. （1） 自动电池数十年不充电也能使用。
   （2） 这样的温度将会引起一些能量变化。
   （3） 计算机只能按指令运行。
   （4） 在未来的战事中，飞机参战将面临导弹的威胁。
   （5） 蓄电池可用来发动汽车。
   （6） 医生决定马上给他动手术。
   （7） 火箭在外层空间的真空状态下飞行。
   （8） 老工人教他怎样开机器。
   （9） 这家公司仅在国内经营。
   （10）新机器运转正常。

Part B Reading

Exercises

1. (1) integer/whole number   (2) fraction
   (3) decimal   (4) square
   (5) cube   (6) raise to any power
   (7) take a square root   (8) take a cube root
   (9) take any root   (10) decimal system
   (11) binary system   (12) geometry
   (13) algebra   (14) probability theory
   (15) group theory

2.（1）人们都说数学是各门学科的基础；而算术，即数的科学，又是数学的基础。

（2）当今普遍采用的另一种进位制是二进制，即二进位计数法。它只用0和1两个数字的组合来表示。

（3）这种进位制完全适用于电的"开—关"脉冲，因此广泛应用于电子计算机中。由于它简单，故常有"懒学生的梦想"之称。

（4）数学的其他分支，如代数和几何，也广泛应用于许多学科中，甚至应用在哲学的某些领域中。数学上更专业化的分支，像概率论和群论，现在的应用范围也在不断地扩大，已从经济学和实验设计扩展到战争和政治领域。

（5）此外，对于新闻记者来说，只要稍有一点统计学知识，就可以避免因为报道失实而将读者引入歧途。而对于普通老百姓来说，则可以利用这方面的知识，来识破总是使之上当的不良企图。

Part C Extended Reading

Exercises

1. (1) differential equation (2) Fourier series
   (3) Laplace transform (4) partial differential equation
   (5) boundary-value problem (6) complex variables/functions of a complex variable
   (7) function (8) limit
   (9) continuity (10) differentiation
   (11) integration (12) infinite series
   (13) improper integral (14) vector analysis
   (15) curriculum (16) graduate study
   (17) undergraduate course (18) literature

2.（1）理工科学生在学完从微积分到微分方程方面的一般课程之后，都面临着决定进一步学什么数学课程的问题。

（2）然而，归根到底，如果研究生们要能阅读本专业的文献资料并在其工作中熟练地应用数学知识，他们之中的大多数人需要透彻地了解应用数学的知识。

（3）我的观点是，想要把物理学或其他工科分支学科作为研究生专业的学生，需要在大学本科时就为其研究生课程打好广阔的数学基础。要做到这一点，他首先应该学一门传统上称之为高等微积分的课程，接着再学一门以高等微积分为基础的数学物理学绪论。

（4）于是，在数学物理学课程中就可以以这门课程为背景，学习相关的应用实例，学生们就能够透彻地了解其中的数学知识。

参考译文

数与数学（Numbers and Mathematics）

人们都说数学是各门学科的基础；而算术，即数的科学，又是数学的基础。数由整数构成，整数则由0、1、2、3、4、5、6、7、8和9这些数字及其任意组合构成。比如247（二百四十七），是由三个数字构成的三位数。小于1的数，有时用分数来表示，但是应用在科学运算上时，却以小数表示，这是因为用小数要比用分数更容易进行各种数学运算。主要的运算有加、减、乘、除、平方、立方或求任意次幂，开平方、开立方或求任意次方根，以及计算出每两个数或每组数之比率或比例。由于这个缘故，所以全世界都把十进制，即十进位计数法，应用在科学上，即使在那些法定度量衡以其他进位制为标准的国家里，也不例外。当今普遍采用的另一种进位制是二进制，即二进位计数法。它只用0和1两个数字的组合来表示。所以在二进制里，2以010表示，3为011，4则以100代之，如此等等。这种进位制完全适用于电的"开—关"脉冲，因此广泛应用于电子计算机中。由于它简单，故常有"懒学生的梦想"之称。

数学的其他分支，如代数和几何，也广泛应用于许多学科中，甚至应用在哲学的某些领域中。数学上更专业化的分支，像概率论和群论，现在的应用范围也在不断地扩大，已从经济学和实验设计扩展到战争和政治领域。最后，统计学的知识是各行各业的科学家分析数据时必须具备的。此外，对于新闻记者来说，只要稍有一点统计学知识，就可以避免因为报道失实而将读者引入歧途。而对于普通老百姓来说，则可以利用这方面的知识，来识破总是使之上当的不良企图。

物理学和工程学中的数学方法
(Mathematical Methods in Physics and Engineering)

理工科学生在学完从微积分到微分方程方面的一般课程之后，都面临着决定进一步学什么数学课程的问题。一些学生完全把数学丢在课程之外，到开始研究生学业后才发现，为了获得解决现代物理学和工程学中正在研究的问题，必须掌握数学基础知识和熟练应用数学基础知识的技巧，他们必须回过头来重新学习大学本科的课程。其他学生由于急于学到基础微积分课程之外的其他方法，于是就学习高等工程数学的若干课程，包括像傅里叶级数、拉普拉斯变换、偏微分方程、边界值问题和复变函数等课程。由于学生们在分析上缺乏扎实的基础，这些课程常常采用启发式的介绍。事实上，一些特定的应用实例会需要这些数学方法，但是在物理学和工程学课程中，依然以同样的启发式方式教给学生这些数学方法。然而，归根到底，如果研究生们要能阅读本专业的文献资料并在其工作中熟练地应用数学知识，他们之中的大多数人需要透彻地了解应用数学的知识。

我的观点是，想要把物理学或其他工科分支学科作为研究生专业的学生，需要在大学本科时就为其研究生课程打好广阔的数学基础。要做到这一点，他首先应该学一门传统上称之为高等微积分的课程，接着再学一门以高等微积分为基础的数学物理学绪论。如果不考虑其实用性，高等微积分就可以成为这样一门课程：它仔细地推导和阐述关于函数、极限、连续性、微分、积分、无穷级数、广义积分，可能还有复变函数等基本概念。于是，在数学物理学课程中就可以以这门课程为背景，学习相关的应用实例，学生们就能够透彻地了解其中的数学知识。本书就是为了满足学习数学物理课程的需要而写的。

该书面向的是业已学完基础微积分、微分方程，若干初等力学课程，以及包括矢量分析在内的上述那些高等微积分课程的学生。了解复变函数对于阅读本书非常有用，但并非不可或缺。即使没有复变量分析的先验知识，也能够理解本书中的大量材料。必要的高等代数知识是从头讲起的，因此，只需掌握基本的大学代数即可阅读本书。

Key and Reference for Translation

Unit Four Physical Science

Part A Lecture

Quiz

1. （1） 这种蒸汽机的**效率**只有 15% 左右。

   （2） 如果一物体重 50 公斤，这就是说，地球对该物体的**吸引力**为 50 公斤。

   （3） 波束可以从一根使用**方便**的短天线上有效地发射出去。

   （4） 指针下的刻度**读数**即为安培数。

   （5） 一门高度发展的自然科学的**特点**是广泛地应用数学。

   （6） 地质学**系统地**研究岩石和矿物。

   （7） 向上的浮力等于向下的重量。

   （8） **不断提高**密闭容器内气体的温度，会使气体内的压力**不断增大**。

   （9） 我们发现，解决这个复杂的问题是**困难的**。

   （10）光学纤维系统传送电话的能力优于普通电缆，因而其**经济性**颇为可观。

Part B Reading

Exercises

1. (1) mechanics   (2) heat
   (3) sound   (4) electricity
   (5) thermodynamics (6) magnetism
   (7) optics   (8) electronics
   (9) chemical physics (10) biophysics
   (11) inertia   (12) solid
   (13) fluid   (14) heat flow
   (15) X-rays   (16) relativity

2. （1） 人们常常把物理学定义为关于物质与能量的科学，它主要涉及物质世界的定律和性质。

（2）综合应用这些领域中所研究的原理，促成了机器时代的到来。像化学物理学和生物物理学这类新术语，就是物理学原理被推广应用的证明，表明其应用范围甚至已扩展到生物机体的研究中。

（3）而具有特殊意义的是涉及力对物体的形状和运动的效应的那些定律，因为这些原理可应用于诸如机器、建筑物和桥梁等设备和结构上。

（4）这类研究促进了能源生产效率的提高、新型高温合金和高温陶瓷材料的研制、接近绝对零度低温环境的产生以及关于物质性质与辐射性质的重要理论的建立。

（5）搞懂电和磁的来源、效应、测量方法及用途等，对于操作工人是很有价值的；有了这些知识就能使他们得心应手地使用当今对提高工作效率和过上舒适生活至关重要的种种电气设备。

（6）光电技术使电视、有线或无线图像传播、有声电影以及许多用电来控制机器设备的装置成为可能。

Part C Extended Reading

Exercises

1. (1) organic chemistry　　　　(2) inorganic chemistry
   (3) analytical chemistry　　　(4) physical chemistry
   (5) biochemistry　　　　　　　(6) electrochemistry
   (7) geochemistry　　　　　　　(8) radiochemistry
   (9) synthetic nutrient　　　　(10) chemotherapy

2. （1）被归类为无机化学的物质主要来自矿物质，而不是来自动植物。

   （2）除自成一门学科外，化学又服务于其他学科和工业生产。

   （3）化学制冷剂使得冷冻食品工业成为可能，这使在其他情况下可能要变坏的大量食物得以保存。化学还能生成合成营养素，但随着世界人口的增长而可耕土地面积有限，仍有很多问题需要解决。

   （4）通过新药开发，医学和化学疗法方面的进步为人类延长了寿命，减轻了病痛。

Key and Reference for Translation

参考译文

物理学（Physics）

人们常常把物理学定义为关于物质与能量的科学，它主要涉及物质世界的定律和性质。这些内容分别成为紧密相关的力学、热学、声学、电学、光学以及原子与核结构等学科中的研究领域。综合应用这些领域中所研究的原理，促成了机器时代的到来。像化学物理学和生物物理学这类新术语，就是物理学原理被推广应用的证明，表明其应用范围甚至已扩展到生物机体的研究中。

力学是物理学最古老的基础分科。这一部分物理学研究涉及惯性、运动、力和能量这类概念。而具有特殊意义的是涉及力对物体的形状和运动的效应的那些定律，因为这些原理可应用于诸如机器、建筑物和桥梁等设备和结构上。力学既研究固体的也研究流体的性质和定律。

热学这门学科研究温度测量原理、温度对材料性质的影响、热流及热力学，即有关热功转换关系的问题。这类研究促进了能源生产效率的提高、新型高温合金和高温陶瓷材料的研制、接近绝对零度低温环境的产生以及关于物质性质与辐射性质的重要理论的建立。

声学研究不仅在音乐和语言方面很重要，而且在通讯和工业方面也很重要。声学与通信工程人员研究的问题涉及声音的产生、传播和吸收等。弄懂声学方面的科学原理对于无线电工程人员有重要意义。工业工程人员对于声音在施工人员中引起的疲劳效应极为关注。

电学和磁学是物理学中的两个研究领域。这两门学科对于电力供应、照明和通讯等技术的迅速发展以及为人类提供便利设施、娱乐和其他领域的研究工具等许许多多的电子装置而言，都特别重要。搞懂电和磁的来源、效应、测量方法及用途等，对于操作工人是很有价值的；有了这些知识就能使他们得心应手地使用当今对提高工作效率和过上舒适生活至关重要的种种电气设备。

光学是物理学中研究光的本质及其传播规律、光反射定律以及光通过棱镜和透镜传播时发生的弯折或称折射现象等的一门学科。其他重要内容还有白色光分解成单色光、光谱的本质及类型、干涉现象和衍射现象以及偏振现象等。测光学则涉及光源发光强度的测量方法以及表面照度的测量方法等。

物理学中一个引人注目的部分通称为现代物理学，包括电子学、原子和核现象、光电学、X 射线、放射学、物质与能量的相互转换、相对论以及与电子管和现代无线电电波有关的现象等。原子的裂变现已为人类提供了实用能源。当今许多常用的设备都是应用现代物理学的一个或几个分科的成果。无线电通信、长途电话通信、扩音器以及电视设备等，只是利用电子管研制出来的许许多多设施中的几个例子而已。光电技术使电视、有线或无线图像传播、有声电影以及许多用电来控制机器设备的装置成为可能。利用 X 射线探测焊接和铸造缺陷，已成为许多工业部门的规范化程序。物理学研究成果的实际应用，现在正以更快的速度继续发展着。

化学与其他学科和行业的关系
（Relationship of Chemistry to Other Sciences and Industry）

化学可大致分为两个主要分支：有机化学和无机化学。有机化学研究含有碳元素的化合物。"有机"一词最初源于研究植物和动物的生物体化学。无机化学涉及所有其他元素以及一些含碳化合物。被归类为无机化学的物质主要来自矿物质，而不是来自动植物。

化学的其他分支，如分析化学、物理化学、生物化学、电化学、地球化学和放射化学，可以看作是两大主要分支的专业领域或辅助领域。化学工程是工程学的一个分支，涉及化工流程的开发、设计和运作。化学工程师通常以化学家的实验室为规模展开，进而将其发展为工业规模的运营。

除自成一门学科外，化学又服务于其他学科和工业生产。化学原理有助于物理学、生物学、农业、工程学、医学、空间研究、海洋学和许多其他学科的研究。化学和物理学是交叉学科，因为两者都是基于物质的属性和表现。生物的过程本质上是化学的。为生物体提供能量的食物代谢是一种化学过程。对蛋白质、激素、酶和核酸的分子结构的了解正在帮助生物学家研究生命细胞的组成、发育和繁殖。

化学在缓解世界日益严重的食物短缺方面发挥着重要作用。通过化肥、农药和改良种子的使用，农业产量得到了提高。化学制冷剂使得冷冻食品工业成为可能，这使在其他情况下可能要变坏的大量食物得以保存。化学还能生成合成营养素，但随着世界人口的增长而可耕土地面积有限，仍有很多问题需要解决。能源需求的增长带来了空气和水污染等棘手的环境问题。化学家和其他科学家正在努力缓解这些问题。

　　通过新药开发，医学和化学疗法方面的进步为人类延长了寿命、减轻了病痛。目前在美国使用的 90% 以上的药品和药物的研发和市场化都发生在近 45 年内。60 年前不为人知的塑料和聚合物工业已经彻底改变了包装和纺织工业，并正在用来生产耐用和有效的建筑材料。化学过程产生的能量可用于供暖、照明和运输。几乎每个行业都依赖于化学制品，如石油、钢铁、橡胶、制药、电子、运输、化妆品、服装、飞机以及电视行业，这个表单还可以列很长。

Unit Five　Earth Science

Part A　Lecture

Quiz

1. （1）碳同氧化合形成多种碳氧化合物。
   （2）当代自然科学正在酝酿着新的重大突破。
   （3）若 A=D，则 A+B=D+B。
   （4）白色的或发亮的表面反射热，而深色的表面则吸收热。
   （5）这种加工方法所需的温度低于熔化该金属所需的温度。
   （6）还要注意，在解决上述问题时，如果使数据对称，就能得到很大程度的简化。
   （7）电动机与发电机的工作原理就是以磁和电这两者之间的关系为基础的。
   （8）这种经验是在极不严格的工作要求和工作环境下取得的，特别是在功耗方面。

2. （1）用于旋转软盘的通常是由负反馈精确控制的电机。
   （2）硬盘系统的盘片由两面覆有一种磁性材料的合金制成。
   （3）另一种可能是，用处于中央部分的不断更新的数据库目录来锁定文件或记录，或同时锁定以避免同时更新或干扰。

（4）1964年，我国爆炸了第一颗原子弹。

（5）太阳中核反应不断向太阳提供能量。

（6）多小型机系统其他的经济方面的优点是更接近用户的实时操作的分布性，更少地依赖于中央设施。

（7）尽管人类最早的祖先也许已经知道或者注意到了寒冷、冰、雪对他们身体以及周围事物的影响——比如他们打猎归来时带回的肉——但是直到中国早期历史阶段，我们才发现早期人类利用自然界中的这种冷冻现象来提高生活水准的迹象，而那时也只是用来冷却饮料。

（8）淬火是把钢加热然后冷却以提高其硬度和抗拉强度，降低韧性，并获得细晶粒组织的过程。这个过程包括把金属加热到临界点即临界温度以上然后迅速予以冷却。

（9）当钢达到这一温度（介于1400华氏度—1600华氏度），如果予以迅速冷却，这一温度变化对获得坚硬的高强度材料是很理想的。如果慢慢冷却，它就会变回到原来的状态。

（10）碳和石墨的性能类似于陶瓷，只有两大性能例外——导电和传热。

Part B　Reading

Exercises

1. (1) volcano　　　　　　　　(2) lava
   (3) crust　　　　　　　　　(4) mineralogy
   (5) petrology　　　　　　　(6) glacier
   (7) coral reef　　　　　　　(8) physical geology
   (9) historical geology　　　(10) machinery of the earth

2.（1）科学工作者试图明确表述这些问题，并依照可以通过观察和实验收集到的有关事实来解决这些问题。

（2）问题不在于我们需要装作什么都懂，而是我们相信自然过程的规律性。经过两三个世纪的科学研究，其结果是：当我们用适当的观察和实验方式，来向大自然提出问题时，我们终于相信，大自然是可以了解的。它会真心实意地答复我们，并且以不断地发现来报答我们。

（3）这样一个雄心勃勃的计划，在力量分配上需要精细地加以划分，而在实践上，把这一学科分成若干分支学科要方便得多。三大分支学科的主要内容是研究地球的岩质结构物，即矿物学和岩石学；研究各种变化据以发生的地质过程或地球机制，即物理地质学；以及研究这种变化的持续变化过程，或地球的历史，即地质史学。

（4）在满足人类需要和工业发展方面，地质学绝不是无足轻重的。

（5）地质学家在野外，带着锤子和地图，过着健康而又精神振奋的生活。地质学家的观察力变得敏锐了，他们对大自然的爱更深厚了，而有所发现的激奋之情，更随时都会迸发出来。

Part C　Extended Reading

Exercises

1. (1) paleoecology　　　　　　(2) comparative anatomy
   (3) archaeology　　　　　　　(4) fossil organism
   (5) body fossil　　　　　　　 (6) trace fossil
   (7) radiometric dating　　　　(8) jigsaw puzzle
   (9) Linnaean taxonomy　　　(10) cladistics
   (11) family tree　　　　　　　(12) molecular phylogenetics

2. （1）古生物学的观察记录可以追溯到公元前 5 世纪，而直到 18 世纪古生物学才基于乔治·居维叶的比较解剖学的成果发展为一门学科，并在 19 世纪得到迅速发展。

   （2）古生物学家运用这些技术发现了很多生命进化的历史，几乎可以一直追溯到地球开始孕育生命时期，即大约 38 亿年前。

   （3）估算这些遗骸所处的年代虽然必要但却很难：有时临近的岩石层会使放射性测定年代方法所提供的绝对年代精确到 0.5%，但更多的时候，古生物学家必须依据相对年代测定方法来解决生物地层学中的"拼图游戏"一类的问题。

   （4）20 世纪的最后 25 年，分子系统发生学得到发展，分子系统发生学通过测量生物基因组内的 DNA 相似程度来研究它们的相近程度。分子系统发生学也可估算物种进化中发生偏离的时间，但这种估算建立在分子钟的基础上，可信度存在争议。

参考译文

地质学的范围（The Scope of Geology）

我们生存的世界提出了种类无穷、引人入胜的种种问题，这些问题激起了我们的惊讶和好奇。科学工作者试图明确表述这些问题，并依照可以通过观察和实验收集到的有关事实来解决这些问题。像"什么？""怎样？""哪里？""何时？"这些问题，要求他们去寻求也许能够给出答案的线索。面对现有的诸多问题，比如关于活火山，我们提出的问题可能有：岩浆的成分是什么？火山如何活动，它的热是怎样产生的？岩浆和气体是从哪里来的？火山在何时初次爆发，下次爆发大致在什么时候？

在所有这些疑问中，"是什么？"这类问题涉及所组成的物质，这可以用化合物和元素来回答。"怎样？"这类问题涉及过程，也就是物质构成、产生或变化的方式。古人把自然过程看成是一些随心所欲的神祇表现其权力的形式；而今天我们把它看作能量作用于物质，或经由物质而表现出来的形式。火山爆发和地震不再是地狱里鬼神行为上反复无常的表现了，而是由于地球内部的热能作用于四周地壳，并经由地壳而产生的。能源存在于地球内部的物质中。当然，在许多方面，我们的知识仍然是不完备的，比方说我们刚才问及关于火山的许多问题中，只有第一个问题是到现在为止可以差强人意地得到解答的。问题不在于我们需要装作什么都懂，而是我们相信自然过程的规律性。经过两三个世纪的科学研究，其结果是：当我们用适当的观察和实验方式来向大自然提出问题时，我们终于相信，大自然是可以了解的。它会真心实意地答复我们，并且以不断地发现来报答我们。

现代地质学的目标是解释从可以在岩石上辨识其最早记录的太古时代直到今天的地球形成全过程。这样一个雄心勃勃的计划在力量分配上需要精细地加以划分，而在实践上，把这一学科分成若干分支学科要方便得多。三大分支学科的主要内容是研究地球的岩质结构物，即矿物学和岩石学；研究各种变化据以发生的地质过程或地球机制，即物理地质学；以及研究这种变化的持续变化过程，或地球的历史，即地质史学。

在满足人类需要和工业发展方面，地质学绝不是无足轻重的。数以千计的地质学家积极地从事勘探和开发地球的矿物资源。他们正在全球搜索煤和石油以及各种有用的金属矿。地质学家还密切关注有关水源的重要问题。许多工程项目，像隧道、运河、码头、水库等，在选择基地和材料的时候，都要听取地质方面的意见。地质学就以这类方式和以许多其他方式来为人类服务。

虽然地质学在研究矿物、岩石和化石的时候，有它独特的实验室方法，但它本质上是一门野外的科学。它吸引着从事这方面工作的人攀大山、临瀑布、履冰川、登火山、循海滩、涉环礁，以搜索有关地球和它那变化得往往令人迷惘的资料。无论是悬崖峭壁间还是采石坑穴里，只要有岩石，就可以观察它的分布和成因，讲出它的历史。地质学家在野外带着锤子和地图，过着健康而又精神振奋的生活。地质学家的观察力变得敏锐了，他们对大自然的爱更深厚了，而有所发现的激奋之情，更随时都会迸发出来。

古生物学（Paleontology）

古生物学是对全新世之前（有时也包括初期）的生物所进行的科学研究。它包括对化石的研究，以此确定生物的进化过程，以及它们彼此、它们和环境的相互影响作用（它们的古生态学）。古生物学的观察记录可以追溯到公元前5世纪，而直到18世纪古生物学才基于乔治·居维叶的比较解剖学的成果发展为一门学科，并在19世纪得到迅速发展。"古生物学"（paleontology）这个词本身源于希腊语，palaios意为"古老的"，on（gen. ontos）指"生物，生命"，logos表示"言语，思想，研究"。

古生物学是生物学和地质学的交叉学科，但和考古学不同，因为它的研究范围排除了具有现代形态的人类。古生物学使用的技术来自各个科学学科，包括生物化学、数学和工程学。古生物学家运用这些技术发现了很多生命进化的历史，几乎可以一直追溯到地球开始孕育生命时期，即大约38亿年前。随着知识积累，古生物学发展了专门的分支学科，其中一些侧重各种化石生物，另一些研究生态和环境史，比如研究古代气候。

实体化石和遗迹化石是古代生物最重要的证据，但在生物体还没有大到足以留下实体化石之前，地球化学方面的证据有助于破解生命进化过程。估算这些遗骸所处的年代虽然必要但却很难：有时临近的岩石层会使放射性测定年代方法所提供的绝对年代精确到0.5%，但更多的时候，古生物学家必须依据相对年代测定方法来解决生物地层学中"拼图游戏"一类的问题。对古代生物进行分类也很难,因为许多生物不适用于通常用于现代生物分类的林奈分类法,因此,古生物学家更多会使用遗传分类学来绘制进化过程的"家族树"。在 20 世纪的最后 25 年,分子系统发生学得到发展,分子系统发生学通过测量生物基因组内的 DNA 相似程度来研究它们的相近程度。分子系统发生学也可估算物种进化中发生偏离的时间,但这种估算建立在分子钟的基础上,可信度存在争议。

Unit Six　　Life Science (I)

Part A　Lecture

Quiz

1.　　几乎每个人都以这样或那样的方式参与设计,甚至在日常生活中也如此,因为出现问题,就必须要解决。设计问题根本不是假设的问题,它有着真实的目的。设计通过采取一定的行动来产生一个最终结果或创造出真实存在的某种事物。在工程学中,不同的人对设计这个词有不同的理解。有些人认为设计师就是拿着制图板草拟齿轮、离合器或其他机器零部件的人。还有人则认为设计是创建一个复杂系统,比如通信网络。在某些工程学领域,设计这一词已经被诸如系统工程或应用决策理论等词所代替。但不论用什么词来描述设计的功用,在工程学中它仍然是使用科学原理和工程工具(数学、计算机、图形学和英语)来制订计划的过程。计划一旦实现,将满足人类的需求。

Part B　Reading

Exercises

1. (1) biologist　　　　　　　　(2) microscopist
   (3) the Renaissance　　　　　(4) microscope
   (5) magnifying lenses　　　　(6) general name

Key and Reference for Translation

(7) the name of genus or group (8) specific name
(9) the name of the species, or kind (10) make contributions to science

2. （1） 世界各地对生物学的发展都做出了诸多贡献。本文将探讨自文艺复兴时期以来的漫长历史阶段里，致力于研究工作的三种不同类型的生物学家。

（2） 范·拉乌文胡克通过透镜观察，认识到整个世界都充满了微生物。

（3） 林奈乌斯的分类法至今仍在使用，有时称为双命名法。

（4） 当达尔文返回英格兰时，他写了一本题为《物种起源》的书，是有关进化的。他的进化论就是根据乘"小猎兔犬"号出游期间的考察结果而形成的。

（5） 范·拉乌文胡克、林奈乌斯和达尔文是生物学历史上三位极其重要的人物。每一位都是对科学做出了重大贡献的一类人中的代表。他们三位对科学都做出了重大的贡献，但他们仅仅是生物学发展史上重要人物中的几位代表而已。

Part C Extended Reading

Exercises

1. (1) cytology (2) histology
   (3) anatomy (4) morphology
   (5) physiology (6) embryology
   (7) genetics (8) paleontology
   (9) taxonomy (10) physiological chemistry
   (11) bioclimatology (12) biogeography
   (13) ecology (14) bioengineering
   (15) biostatistics (16) bioenergetics
   (17) microbiology (18) psychology
   (19) bacteriology (20) anthropology

2. （1） 生物学是研究生物的科学，主要包括研究植物的植物学和研究动物的动物学。

（2）在史前洞穴艺术中，我们可以看到早期人类观察自然的证据。

（3）16世纪显微镜的发明也大大促进了生物学的发展，拓宽并深化了其范围，并且创立了微生物学（研究生命的微观形式）和显微镜学（对活细胞的微观研究）。

（4）植物学是研究植物的科学，在人类知识史上具有特殊的地位。几千年来，植物学是人类唯一超越了模糊认识的领域。

（5）对他们而言，植物学本身没有名称，甚至可能根本不被认为是"知识"的一个特殊分支。

（6）不幸的是，工业化程度越高，我们与植物的直接接触就越遥远，对植物学的了解也更模糊。

（7）大约一万年前的新石器时代，生活在中东的祖先们发现某些草可以收割，而它们的种子种植后可以在下一季收获更大，这是人类在与植物产生的新联系中迈出的重要的第一步。

（8）从那以后，人类逐渐以少数种植的农产品为生，而不是从品类繁多的野生植物中获取，其结果则是，在千万年间对野生植物的经验和亲密接触所积累的知识，也就随之逐渐衰减了。

（9）当生理学与化学和其他物理学科相结合时，人们对生理过程的了解大大增加了。

（10）现代动物学不仅研究细胞、细胞的组成和功能，还致力于拓展细胞学、生理学和生物化学的知识，并且探索了心理学、人类学和生态学等领域。

参考译文

生物学的历史（History of Biology）

世界各地对生物学的发展都做出了诸多贡献。本文将探讨自文艺复兴时期以来的漫长历史阶段里，致力于研究工作的三种不同类型的生物学家。他们对生物学的历史发展乃至现代科学，都做出了重大的贡献。本文将探讨这三类生物学家中每一类的一位代表人物所做的工作。

第一类生物学家，也即这里要研究的最早的一类生物学家，是17世纪的显微镜学家。这些人靠显微镜从事研究工作。他们制作显微镜并不断改进显微镜,以便应用于科研工作中。荷兰显微镜学家安东尼·范·拉乌文胡克(1632—1723)就是这一类生物学家中重要人物的代表。范·拉乌文胡克对于改进用来制作显微镜的透镜感兴趣。他制作过几台显微镜,并且借助他自己制作的放大镜(凸透镜)对许多东西进行了仔细观察。范·拉乌文胡克通过透镜观察,认识到整个世界都充满了微生物。当时大多数人并不知道这些小生命的存在。现在显微镜仍然是科学研究极重要的工具。

　　再一类是18世纪的生物学家,他们的工作是把人类的科学知识进行系统的整理。他们试图把由众多科学家发现的全部知识都加以系统化,以便使大家在谈论科学发现时,都能使用相同的分类法。其中一种分类法就是由一位名叫卡儿·冯·林奈乌斯的瑞典科学家(1707—1778)创立的。他用一种非常有用的方法将植物、动物和矿物进行了分类。他主张给每一种植物、动物和矿物取一个由两部分构成的拉丁名称。其第一部分是通称,它指明一种植物、动物或矿物属于哪一大类,这是类属名称。而其第二部分则是种名,是物种名称或品种名称,它指明是哪一种植物、动物或矿物。

　　这种分类法在科学家中极为流行,而且至今仍在使用。之所以流行有几个理由。第一,这种方法既简单又明了。第二,林奈乌斯在其分类法中用了拉丁文,而在当时几乎每个科学家都懂拉丁文。因此,凡是会拉丁文的人都不需要学习任何其他的语言。此外,所取的两部分名称每个都是一条简洁的说明,相当容易记忆。林奈乌斯的分类法至今仍在使用,有时称为双命名法。

　　还有一类生物学家所做的大多数工作是在19世纪。这些科学家从当时的全球考察热中受益匪浅。他们作为考察者、收集家进行过多次远航,其工作就是研究新陆地的动植物。最著名的探险家和考察家之一就是伟大的英国生物学家查尔斯·达尔文。他于1809年至1882年间生活在英格兰。

　　作为一名探险家,达尔文没有在英格兰家里度过他的一生。在19世纪30年代初期,他离开英格兰达五年之久,乘一艘名叫"小猎兔犬"的轮船出游。那是一次著名的旅行。对于船上的其他人而言,那次旅行的目的是绘制地图,到南美洲探险。他们还曾计划环绕地球航行。而对于达尔文来说,目的却是不同的。他从南美洲和临近海域采集了许许多多动植物标本,还写下了许多对于探险期间所发现的生物进行考察的结果。当达尔文返回英格兰时,他写了一本题为《物种起源》的书,是有关进化的。他的进化论就是根据乘"小猎兔犬"号出游期间的考察结果而形成的。

范·拉乌文胡克、林奈乌斯和达尔文是生物学历史上三位极其重要的人物，每一位都是对科学做出了重大贡献的一类人中的代表。他们三位对科学都做出了重大的贡献，但他们仅仅是生物学发展史上重要人物中的几位代表而已。

生物学分支（Branches of Biology）

生物学简介

生物学是研究生物的科学，主要包括研究植物的植物学和研究动物的动物学。每一类还可以细分为细胞学（细胞研究）、组织学（组织研究）、解剖学或形态学、生理学和胚胎学（个体动物和植物的胚胎发育研究）。生物学研究还包括遗传学、进化论、古生物学、分类学或系统学（分类研究）。生物学研究中如生物化学（生理化学）、生物物理学（生命过程物理学）、生物气候学和生物地理学（生态学）、生物工程学（人造器官设计）、生物统计学、生物能量学和生物数学，这些领域也会运用其他学科的研究方法和态度。在史前洞穴艺术中，我们可以看到早期人类观察自然的证据。古希腊人就已经开始发展生物学概念。亚里士多德的生物学著作包括他对大量动物的观察和分类。16世纪显微镜的发明也大大促进了生物学的发展，拓宽并深化了其范围，并且创立了微生物学（研究生命的微观形式）和显微镜学（对活细胞的微观研究）。很多人对科学做出了贡献，其中有克劳德·伯纳德、居维叶、达尔文、赫胥黎、拉马克、林奈乌斯、孟德尔和巴斯德。

植物学

植物学是研究植物的科学，在人类知识史上具有特殊的地位。几千年来，植物学是人类唯一超越了模糊认识的领域。虽然今天我们还无法知道石器时代的祖先们对植物了解的程度，但是从我们对仍然存在的前工业社会的观察来看，祖先们对植物及其特性的详细了解必须是久远的。这是合乎逻辑的。植物是所有生物的食物金字塔的基础，甚至对于其他植物也是这样。植物一直对造福人类极为重要，不仅仅是食物，在衣服、武器、工具、染料、药物、住所等许多其他方面也可体现。如今生活在亚马孙丛林中的部落可以识别数百种植物，并了解每一种植物的许多特性。对他们而言,植物学本身没有名称，甚至可能根本不被认为是"知识"的一个特殊分支。

不幸的是，工业化程度越高，我们与植物的直接接触就越遥远，对植物学的了解也更模糊。然而，大家在不知不觉中掌握了大量的植物学知识，几乎没人不认识玫瑰、苹果或兰花。大约一万年前的新石器时代，生活在中东的祖先们发现某些草可以收割，而它们的种子种植后可以在下一季收获更大，这是人类在与植物产生的新联系中迈出的重要的第一步。谷物被发现后，种植作物这一农业奇迹由此产生。从那以后，人类逐渐以少数种植的农产品为生，而不是从品类繁多的野生植物中获取，其结果则是，在千万年间对野生植物的经验和亲密接触所积累的知识也就随之逐渐衰减了。

动物学

动物学是研究动物生命的科学。从很早开始，动物对人类而言就格外重要。洞穴艺术展现了动物对史前人类来说既实际又神秘。早期对动物的分类是基于外貌、栖息地和经济用途。尽管希波克拉底和亚里士多德对他们那个时代的科学思想方面做了很多贡献，但在罗马人的统治下，系统的调查研究减少了，并且继盖伦做出显著贡献后，几乎在整个中世纪时期停滞了（阿拉伯医师除外）。随着文艺复兴的到来，人类重新开始直接观察自然，影响很大的是维萨里的解剖学和哈维关于血液循环的论证。显微镜的发明和实验技术的利用将动物学扩大为一个领域，并建立了许多分支，如细胞学和组织学。胚胎学和形态学的研究揭示了动物生长的本质以及它们的生物学关系。林奈乌斯设计的双命名法可以表现这些关系，他是第一个使其保持一致并系统应用它的人。古生物学是研究化石生物的科学，由居维叶于1812年创立。当生理学与化学和其他物理学科相结合时，人们对生理过程的了解大大增加了。1839年，细胞学说建立起来，30年后人们接受了细胞质是生命组成部分的说法，这些都推动了遗传学的发展。拉马克、孟德尔和达尔文提出了革命性的科学思想概念。他们的进化论和遗传物理基础理论促使人们对所有生命过程和所有生物体的关系进行研究。巴斯德和科赫的经典著作开创了细菌学这一领域。现代动物学不仅研究细胞、细胞的组成和功能，还致力于拓展细胞学、生理学和生物化学的知识，并且探索了心理学、人类学和生态学等领域。

Unit Seven　Life Science (II)

Part A　Lecture

Quiz

1. （1）物质具有一定的特征或特性，所以我们能够很容易地识别各种物质。

   （2）分子具有很好的弹性，所以它们碰撞后没有能量损失。

   （3）物体具有的热量是由几个因素决定的，温度只是因素之一。

   （4）一个典型的例子就是普通的温度计，我们在日常生活中对它是很熟悉的。

   （5）他们想出一种办法，采用这种办法，产量现已迅速提高。

   （6）用来做功的作用物的价值取决于它在一定时间内能做多少功。

   （7）这家工厂生产附有一些精密仪器的机床。

   （8）使用哪一种润滑剂主要依轴承的转速而定。

   （9）现在人们已经懂得，如果食物中缺少了某些重要成分，即使其中不含有任何有害物质，也会引起严重疾病。

Part B　Reading

Exercises

1. (1) double-helical nature of DNA　　(2) molecular biology of the gene
   (3) human genome　　(4) genetic roadmap
   (5) genetic code　　(6) the Primacy of Genes
   (7) schizophrenia-prone gene　　(8) schizophrenia-inducing experience

2. （1）在几乎所有罗列20世纪最有影响力人物的名单中，都有詹姆斯·沃森和弗朗西斯·克里克，这并不令人惊奇，他们理应和丘吉尔、甘地和爱因斯坦并肩而立。

   （2）与此同时，许多人对这个科学里程碑的前景感到害怕也是不足为奇的。这一发现无疑是革命性的，但是个中却蕴含着《勇敢的新世界》

式的令人恐慌的色彩，它给人们提出了一些有关我们将怎样认识我们自己的基本问题。

（3）如果你持有这种错误的基因观，哇，那你就认可了某种可怕的东西，像某人具有病态的寻衅行为，或某人有精神分裂症的倾向。总之，从一开始孕育，人的一切就已注定了。

（4）但实际上几乎没有基因是以这种方式运作的。的确，基因和环境是相互作用的：后天的培育会加强先天的禀性。

（5）如果你认为基因是自发作用的，那么答案就是它们本来就知道。没有人告诉基因该做什么，它们自主行事——自发地工作，自发地停止。

（6）还有就是：在期末考试期间痛苦的、充满压力的一天将会激活一个大学生体内的某种基因，这种基因能抑制人体的免疫系统，通常会导致感冒或更严重的情况。

（7）基因学研究从来没有这样包罗万象，以至于可以涵盖从医学到社会学的每门学科。然而，科学解释的基因信息越多，我们就越了解环境的重要性。这点同样适用于现实生活：基因确实非常重要，但它并不是生命的全部。

Part C Extended Reading

Exercises

1. (1) legume   (2) additive
   (3) mayonnaise   (4) margarine
   (5) cooking oil   (6) salad dressing
   (7) coffee creamer   (8) cereal
   (9) shortening   (10) fertilizer
   (11) pesticide   (12) irrigation
   (13) high-yield crop (14) herbicide
   (15) allergic reaction

2. （1）虽然还不能说每一口食物都是经过基因工程改造过的，但是几乎每一个美国人都吃过通过基因工程改造过的食物。

（2）这一次，不是用近亲植物杂交法，而是插入基因。

（3）这种技术显示出了极大的潜力，"应用生物技术将会产生我们根本想象不到的各种可能，即使我们的想象力长了翅膀也想象不出来"。

（4）科学家还怀疑，带有插入基因的食物改善了它们某一方面的性质的同时，会不会破坏了其他方面的性质？其一是这些基因可能引起人们对本来不过敏的食物产生过敏反应。

参考译文

基因并非包含一切（It's not "All in the Genes"）

在几乎所有罗列20世纪最有影响力人物的名单中，都有詹姆斯·沃森和弗朗西斯·克里克，这并不令人惊奇，他们理应和丘吉尔、甘地和爱因斯坦并肩而立。在识别DNA双螺旋结构性质的过程中，沃森和克里克为人类理解基因分子生物学铺平了道路，这是战后时期主要的科学成就。为人类基因组排序将意味着由这两位天才所发动的革命近乎结束。

与此同时，许多人对这个科学里程碑的前景感到害怕也是不足为奇的。这一发现无疑是革命性的，但是个中却蕴含着《勇敢的新世界》式的令人恐慌的色彩，它给人们提出了一些有关我们将怎样认识我们自己的基本问题。它是否意味着我们的行为、思想和情感只是我们基因的总和，而且科学家是否能用一张基因图谱就可以计算出那个总和是什么？那么，我们是谁呢？我们所珍视的个性意识和自由意志将会怎么样？知道我们自己的遗传密码是否就意味着知道我们注定的命运？

我并没有这样的害怕，让我来解释为什么。这种焦虑的症结在于人们有基因至上的想法。这就是当你解释生物学上的一些复杂问题时（例如为什么一些特殊的鸟会迁徙到南方过冬，或者为什么那个人会患精神分裂症）所持有的观点，答案就在于了解构成这些现象的基本因素——基因至上者便将这些基本因素解释为起决定作用的基因。在这种宿命论的观点看来，由基因释放出的蛋白质将导致或控制行为。如果你持有这种错误的基因观，哇，那你就认可了某种可怕的东西，像某人具有病态的寻衅行为，或某人有精神分裂症的倾向。总之，从一开始孕育，人的一切就已注定了。

但实际上几乎没有基因是以这种方式运作的。的确，基因和环境是相互作用的：后天的培育会加强先天的禀性。例如，研究者指出"携带有精神分裂基因"意味着有 50% 的风险会得这种疾病，而不是肯定会得。只有当人们同时具有易患精神分裂症的基因和具有诱发精神分裂症的经历时，才会得病。某种特定的基因能够有不同的作用，这取决于环境。基因有遗传的脆弱性而并非必然性。

基因至上论还认为基因可以自发作用。然而它们怎么知道什么时候开始或结束特定蛋白质的合成呢？如果你认为基因是自发作用的，那么答案就是它们本来就知道。没有人告诉基因该做什么，它们自主行事——自发地工作，自发地停止。

可是，这种观点也远远算不上准确。结果证明，在长度惊人的 DNA 序列中，事实上只有极少一部分字母组成了构成基因的词汇并充当蛋白质的密码。DNA 中 95% 以上的部分是没有密码的。大多数的 DNA 仅仅构成了调节基因活动的开关。这正如你有一本 100 页的书，其中的 95 页全是为其余 5 页的阅读作说明和指导的。所以，基因不能单独决定什么时候合成蛋白质。它们是执行来自另外某个地方的指令。

那么用什么来调节这些开关呢？在一些情况下，用来自于细胞其他部分的化学信使；在另一些情况下，用来自于身体其他细胞的信使（这就是许多激素作用的方式）。而且，非常关键的是，还有一些情况下，环境因素会打开或关闭基因。举一个粗浅的例子，一些致癌物通过进入细胞内部起作用，并把自己同一个 DNA 开关结合起来，开启那些引发无限制生长的基因而引起癌症。另外的一个例子就是：一只雌性老鼠舔舐自己的孩子为其梳理毛发的行为将会引发一系列的事件，这些事件最终开启幼鼠体内与成长相关的基因。再就是处于发情期的一只雌性哺乳动物的气味将会激活特定灵长类动物体内与繁殖相关的基因。还有就是：在期末考试期间痛苦的、充满压力的一天将会激活一个大学生体内的某种基因，这种基因能抑制人体的免疫系统，通常会导致感冒或更严重的情况。

你不能将基因与控制其开关的环境分离开来。你也不能将基因的作用与蛋白质作用的环境分离开来。基因学研究从来没有这样包罗万象，以至于可以涵盖从医学到社会学的每门学科。然而，科学解释的基因信息越多，我们就越了解环境的重要性。这点同样适用于现实生活：基因确实非常重要，但它并不是生命的全部。

转基因食品（Genetically Modified Foods）

转基因食品（经基因改造加工过的食物）已端上了美国人的餐桌。这些食品安全吗？

虽然还不能说每一口食物都是经过基因工程改造过的，但是几乎每一个美国人都吃过通过基因工程改造过的食物。

就拿大豆来说，美国大豆协会报道，去年这种作物大约有55%在性状上都经基因工程改造过。虽然很少有美国人坐下来弄一盘大豆当晚餐，但是这些豆科植物已通过秘密途径作为添加剂加到多种食物中，如蛋黄酱、人造黄油、食用油、沙拉酱、咖啡伴侣、啤酒、谷类食品、糖果、起酥油等。

美国处在第二次绿色革命的边缘。在20世纪后半叶的仅仅30年里，第一次绿色革命就把世界食物产量增至3倍。科学家通过使用理想性状的相关植物杂交来增加作物产量。然后农场主用化肥、杀虫剂、灌溉等方法使高产作物生长旺盛。

既然这些庄稼的产量已平稳，科学家便又开始寻找生物技术以继续增加植物的产量。这一次，不是用近亲植物杂交法，而是插入基因。例如，比目鱼基因可以帮助像西红柿和草莓之类的普通植物抗寒。研究人员还把细菌基因插入到玉米和大豆这些植物中，以便更好地预防病虫害或提高这些植物对某些除草剂的免疫力。

这种技术显示出了极大的潜力，"应用生物技术将会产生我们根本想象不到的各种可能，即使我们的想象力长了翅膀也想象不出来"。

对农场主来说，基因工程的庄稼能抗寒、抗旱，或抵御其他不利的天气条件，可做到既少花钱又少出力还能提高作物产量。对消费者来说，还意味着食物更便宜。这种庄稼还可以让发展中国家的人民有饭吃，有更好的营养，例如像非洲这样干旱的地方。

把基因序列转入像玉米这样的庄稼，可以使它们自己分泌杀虫物质，这对环境有利；农场主再也用不着喷洒这类杀虫剂了。将来，我们理想的食物能预防癌症，水果和蔬菜能释放疫苗，庄稼甚至可以给自己施肥。

但是，一旦这种庄稼开始侵入我们的食品供应中，环境组织和消费者组织也就会开始质疑：这种庄稼对环境和人类健康潜在的危害我们是否做了充分的研究。比如说去年，生物学家首次发现证据，在野外播种基因改造过的玉米可能杀死了蝴蝶，因为蝴蝶吸食了这种玉米的花粉。科学家还怀疑，带有插入基因的食物改善了它们某一方面的性质的同时，会不会破坏了其他方面的性质？其一是这些基因可能引起人们对本来不过敏的食物产生过敏反应。

例如，20世纪90年代中期，在依阿华州的得梅因，先锋高杂粮国际公司想要通过引入巴西坚果基因提高大豆的品质。但检测后发现，这种大豆引发那些对巴西坚果过敏的人的荨麻疹反应。于是这家公司就放弃了这个项目。由于对坚果过敏的现象很普遍，因此检测转基因大豆的反应变得很有意义。

但是改变作物的基因需要进行反复试验。当一个科学家把一个新基因插入某种庄稼，他会得到几种不同类型的种子。有些插进了错误的位置，有些根本未插入，最终研究人员可以获得他想要的种子。然而，要是这个基因将自己微妙地插入该植物DNA某个部位，又该如何呢？

Unit Eight　Aeronautics

Part A　Lecture

Quiz

1. （1）最常见的关节炎有骨关节炎和风湿性关节炎两种。前者的发病率随年龄的增长而上升，而后者则可见于任何年龄，甚至包括幼儿期。
   （2）就别的金属而言，主要是铅、汞和锡，当达到某一温度时，其电阻下降为0。这个温度对于上述三种不同的金属，各不相同且都为0度以下。
   （3）他们得对日益增多的一系列的结果做出解释。这些结果不仅同人类过去对自然的了解相悖，而且刚开始时彼此之间也很不一致。
   （4）工业家长期以来一直依靠物理冶金学家来解决一些迫切的问题——即使这可能需要物理冶金学家另起炉灶去掌握金属和合金的基本特性。而现在物理冶金学家却想加入到金属物理学家的行列，尽管他所受过的训练不利于他从事新的职业。
   （5）在所谓的电镀工艺中也应用了这一原理。通过电镀在母体金属上可沉积一层很薄的诸如铬或锡一类金属面层；进而这层金属面层可牢固地附着在母体金属上。

（6）铝的潜力全部开发出来耗费了从1890年到1949年大约50年的时间，而以目前物理学家和工程师们思维和规划更新的速度，要把新金属的潜力充分挖掘出来则再也不需要50年那么漫长的时间了。

（7）然而，德国海军陆战队却得出结论说，用以攻击防卫严密的水面目标的最佳武器是优良的老式铁制炸弹。这对上述观点的支持者来说，可能会感到吃惊。

（8）物质不会由无到有，也不会从有到无；化学反应物的质量总和与化学反应后生成物的质量总和是恒等的，这就是质量守恒定律。我们通过科学实践认识了上述规律。

（9）发言人说，人们看到那艘轻型巡洋舰和那艘护卫舰正由一艘油轮伴随向南驶去，显然是要驶往中国南海地区和其他舰只会合。

（10）计算机语言有低级的，也有高级的。前者比较详细，很接近于特定计算机直接能懂的语言；后者比较复杂，适用范围广，能自动为多种计算机所接受。

（11）20世纪初，技术条件产生了深刻的变化，有了极大的改进。但在当时的技术条件下，燃料、材料、安全着陆、推进装置，特别是电子计算等一系列的问题都过于复杂，难以解决，导致没有登月运载工具。因此直到20世纪后期才揭开月球的秘密。

（12）有的河的水量在一年中的某些季节较少，而在某些季节里却多得多；有的河利用其动力大量发电；有的河所负担的运输量很大。

（13）这份文件的目的，是向这次会议提供一份关于早先所做决议的参考资料，以便有助于对当前问题的讨论，而不在业已讨论过的问题上浪费不必要的时间。

Part B Reading

Exercises

1. (1) aircraft engineering (2) a matter of life or death
   (3) power of engine   (4) fuel consumption

Key and Reference for Translation

(5) radio navigational instrument (6) passengers seat
(7) freight room (8) normal load
(9) stress man (10) accidental damage
(11) ground strength test (12) fatigue strength
(13) certificate of airworthiness (14) flying controls
(15) electrical equipment (16) fire precaution
(17) below freezing point

2.（1）飞机工程中有两个主要难题：一是要使每个部件尽可能可靠；二是要把一切都制造得尽可能地轻。飞机在空中飞行，出了故障又不能停飞，因此，飞机的性能必须绝对可靠，这也许是一个生死攸关的问题。

（2）在发动机功率已定，并且耗油率也因此确定的情况下，要使飞机飞行，飞机的总重量实际上要受到一定的限制。

（3）他要考虑可能加在部件上的任何异常应力，作为预防制造中可能产生的误差、意外损坏等的措施。

（4）地面强度试验有两种：一种是求机翼、机尾等组件在达到最大载荷而被破坏之前的抗载荷强度。或译为：必须做两种地面强度试验。第一种是求出机翼、机尾等处负荷的阻力，这种测试一直进行到负荷达到最大值导致部件毁坏为止。

（5）作为单独的一个载荷来说，每一次（载荷量）都可能远远低于结构所能承受的限度，但是多次重复以后即可导致飞机毁坏。

Part C Extended Reading

Exercises

1. (1) fuel valve (2) electrical generator
   (3) hydraulic pump (4) auto-pilot
   (5) navigational system (6) commercially advantageous to
   (7) fuselage (8) punched-card

2.（1）飞机的制造同它的设计一样是非常繁杂的，它牵涉到复杂且贵重机器的使用，并且要求人的组织计划能力达到最高限度。

（2）一架新飞机的建造是如此复杂的原因之一就是为了有助于确保新飞机必须比它的上一代在性能上有一定的改善，许多系统部件如燃油阀、油泵、发动机和阀门开关，以及液压泵和液压制动器均需用新的，而一概不用"现成的"部件。

（3）某些附件的开发，如自动导航和领航系统的电子设备极其复杂，所要求的设备也与飞机机身的制造商所用的设备一样复杂和昂贵。

（4）机翼和机身的建造是从主翼梁和翼肋以及机身骨架的夹紧支撑开始的，余下的结构则在此基础上建造。

（5）利用自动控制（加工）有许多优点，包括节省了切割机人工定位因效率差所占去的时间，定位精度也大大提高了，误差仅为一英寸的万分之一。

参考译文

飞机制造者（The Plane Makers）

飞机工程中有两个主要难题：一是要使每个部件尽可能可靠；二是要把一切都制造得尽可能地轻。飞机在空中飞行，出了故障又不能停飞，因此，飞机的性能必须绝对可靠，这也许是一个生死攸关的问题。

在发动机功率已定，并且耗油率也因此确定的情况下，要使飞机飞行，飞机的总重量实际上要受到一定的限制。在这总量之内，要求尽可能多地来安排燃油、无线电导航仪器、旅客座位或货舱，当然还有旅客和货物本身。因此，在安全和效率允许的情况下，飞机的结构必须尽可能地做到小且轻。设计师必须计算出每一部件所能承受的正常载荷。这种专家称为"应力计算专家"。他要考虑可能加在部件上的任何异常应力，作为预防制造中可能产生的误差、意外损坏等的措施。

应力计算专家将计算结果交给部件的设计人员，部件设计人员必须使部件达到应力计算专家所要求的强度。为了证明部件的强度符合设计人员的要求，总要对一两个试样进行试验。每一个部件都要进行单独的试验，然后再对整个组件（如整个机翼）进行试验，最后则对整架飞机进行试验。当制造一种新型飞机的时候，在最先制造出来的三架飞机中，通常只有一架飞机能飞行。两架飞机会在地面上进行的结构试验中被破坏。第三架飞机则在空中进行试验。

地面强度试验有两种：一种是求机翼、机尾等组件在达到最大载荷而被毁坏之前的抗载荷强度；另一种是疲劳强度试验。要施加无数次较小的载荷。作为单独的一个载荷来说，每一次（载荷量）都可能远远低于结构所能承受的限度，但是多次重复以后即可导致飞机毁坏。

飞机通过这一切试验之后，就可以获得政府颁发的适航证。如果没有这种证书，除了试飞之外，一切飞行都是不合法的。

把结构造得具有足够的强度是很难的。同样，把工件造得可靠也很难。飞行控制机构、电力设备、防火设施等不仅要做到重量轻，而且应该既能在冰点以下的低温环境，也能在热带地区的酷热机场进行工作。

为了解决这一切问题，飞机工业配备了大量的研究人员和精心设计的实验室和测试场，同时能对相应重量提供最佳强度的新型材料也在不断测试之中。

飞机是怎样制造出来的（How Aircraft Are Built）

建造新飞机的开始阶段称为"金属加工"阶段，这一表述非常恰当，亦如下文所述。飞机的制造同它的设计一样是非常繁杂的，它牵涉到复杂且贵重机器的使用，并且要求人的组织计划能力达到最高限度。有关的组织和计划工作很可能是不常被考虑的事，然而少了这些，永远就不会有新飞机飞向蓝天。

一架新飞机的建造是如此复杂的原因之一就是为了有助于确保新飞机必须比它的上一代在性能上有一定的改善，许多系统部件如燃油阀、油泵、发电机和阀门开关，以及液压泵和液压制动器均需用新的，而一概不用"现成的"部件。这里要用到上千个附件，这些附件的设计与开发，通常由不同领域的专家们完成。某些附件的开发，如自动导航和领航系统的电子设备极其复杂，所要求的设备也与飞机机身的制造商所用的设备一样复杂和昂贵。当飞机设计师很可能自己也不确切知道他需要什么时，预测并说明设计生产如此这般的一个部件将用"两年四个月"的时间是一件不那么容易的事情。然而不论复杂程度如何，每个部件在具体的时间必须满足需要以符合全部的生产计划。这种生产计划限定的时间总是被尽可能的压缩，因为一旦决定一架新飞机"上马"，尽早生产出来并飞行总是有利可图的。

就机身的建造而言，工作自然是从那些设计制作时间较长以及在生产计划中首先需要的部件开始，许多关键的安装接头由铸件、挤压件、锻件经过机床加工而成，铸坯和坯段都是预先订购的。金属的切割和机床加工通常是从机身骨架和翼梁的建造开始的。

  机翼和机身的建造是从主翼梁和翼肋以及机身骨架的夹紧支撑开始的，余下的结构则在此基础上建造。

  传统的机身和机翼的制造方法是把金属薄板固定在铆接的桁条上，并把它用铆钉和螺栓与机身骨架和翼梁固定起来。

  要提高飞机的性能和增加飞机的尺寸，就必须发展新的制造技术。加工机翼蒙皮板就是新技术之一。在制造加工过程中，舱盖切口周围的各种成形件、桁条、加强筋与蒙皮进行整体成形加工，蒙皮通常加工成不等的厚度，在载荷较小的地方加工得薄一些，以尽量去除每一点重量。

  这种制造方法消除了铆接孔，解决了整体密封燃油箱的问题和承压机身的问题。更重要的是，它具备了良好的耐久性能。这是因为铆接位置载荷高度集中的问题不再存在，还因为大半径的交切点可以用机床加工，以"平摊"载荷分布，同样强度的整体加工蒙皮壁板的重量要比装配加工的轻得多。

  利用自动控制（加工）有许多优点，包括节省了切割机人工定位因效率差所占去的时间，定位精度也大大提高了，误差仅为一英寸的万分之一。此外，任何时候切割机的位置均可重复调整，无须复杂和昂贵的钻孔模，若设计上出现改动，需要的不是一套新的钻孔模而是一套新的穿孔卡指令。

Unit Nine Astronautics

Part A Lecture

Quiz

1. 该段落的主题是 transmitter。

  参考译文：发射机通常由几个部分组成，它是发射无线电波的设备。无线通讯发射机的用途是用无线电来发送信息。为了用无线电发送信息，必须产生高频信号，因为只有在高频时才能发射无线电波。

2. （1） 有两种制约人智商的因素。第一种因素是人生来具有的大脑。人的大脑差异很大，一些人比另一些人强。但是，不管一个人生来大脑有多好，如果没有学习机会，也会出现智力低下的情况，所以另一种因素是发生在人身上的事、所经历的环境。如果一个人的环境是畸形的，很可能他的大脑就不会良性发育，也不会达到本该有的智商水平。

（2） 伦琴在他的发现报告中称这种射线为 X 射线，因为 X 通常代表一些未知的事物。其他科学家为了纪念这位第一个发现 X 射线的人，把它称作伦琴射线，但现在人们还是常用 X 射线这个名字。

（3） 沥青结合砖成型后冷却入库，或径送轻烧窑轻烧。这种砖经过轻烧后，不少性能都得到改善。

（4） 磁悬浮列车至少有两节车厢。每节平均设座位约 90 个。根据实际情况及客流量，列车可多达 10 节（首尾两节为机车和守车，中间 8 节为车厢）。

（5） 因此，电子学是各种电信系统的基础。含有某种极低气压的气体的无线电真空管，现今正被称为晶体管的固态半导体装置所取代；因为同气体真空管相比，晶体管体积小得多，而且寿命也长。自从第二次世界大战以来，制造与开发固态电子产品的整个新型工业已经建立起来。这一领域的一个新的分支，促进了计算机的发展，因为计算机里装有数目庞大的晶体管、小型磁芯和其他新型固态电子装置。

Part B Reading

Exercises

1. (1) deep space probe　　(2) lunar probe
   (3) rocket vehicle　　(4) sounding rocket
   (5) elliptical path　　(6) artificial satellite
   (7) ultraviolet light　　(8) infrared ray
   (9) X-ray　　(10) radio wave

2. （1） 第二次世界大战以来，特别是最近几年，在运用大功率火箭将仪器送入远离地球的高空、发射人造卫星和外层空间探测器方面,人们做出了越来越多的努力。

(2) 主要原因之一是,我们的大气层尽管总体上来说是有利于生命的,但是,我们只能在一个非常有限的光域内看见宇宙,实际上,几乎完全被限制在可见光和相当有限的无线电波范围之内。

(3) 我们不知道这些仪器会记录些什么,如果知道的话,就不值得找这么多麻烦了。但是这个天文学研究领域有待于以后几代人进行研究。

(4) 这只是科学研究中诸多新的可能性之一。在研究地球外部大气层、气象学和地球与行星间的空间等过程中,火箭运载工具的发展开辟了这些新的可能性。

(5) 如果卫星轨道特别长,超过地球半径(4,000英里)数倍时,我们就采用外层空间探测器。

(6) 人造卫星的出现并没有使垂直探测火箭过时,外层空间探测器和月球探测器与人造卫星的关系更是如此。

(7) 科学家也不关心这些运载工具是否载人,因为仪器可以自动操作并将其记录的数据转换成无线电信号,从数百万英里以外送回到地球。

Part C Extended Reading

Exercises

1. (1) 自古以来,人们就梦想着离开我们的星球,探索外部世界。20世纪下半叶,梦想成为现实。1957年第一颗人造卫星的发射拉开了太空时代的序幕。1961年,人类首次进入太空。从那以后,太空探索持续时间更长,宇航员甚至连续数月生活在环绕空间站飞行的飞船上。

(2) 从火箭到通讯设备再到计算机,这些探索活动促进了新技术的发展。宇宙飞船的研究得到了大量关于太阳系、银河系和宇宙的科学发现,并且为人类了解地球及相邻星球提供了新的视野。

(3) 由于太空的真空状态,载人宇宙飞船需要自带空气。此外,还有其他致命的危险,如太阳和宇宙的辐射;可能会刺穿宇宙飞船船身或戳破宇航员的压力服的微陨石群(小的岩石和尘埃);从寒冷的黑夜到火辣辣的阳光直射引起的极冷到骤热的温度变化；等等。

（4）目前，科学家们主要致力于两个方面的工作：一是在执行长达一年或更长时间（需要到达附近行星的时间）的太空任务时，保持宇航员的健康状态；二是降低发送卫星到轨道的成本。

参考译文

宇宙科学探索（The Scientific Exploration of Space）

第二次世界大战以来，特别是最近几年，在运用大功率火箭将仪器送入远离地球的高空、发射人造卫星和外层空间探测器方面，人们做出了越来越多的努力。我们已经指出了人们在地球表面已经做了多少，还能做多少。那么，为什么还要把注意力集中到火箭的运用上呢？

主要原因之一是，我们的大气层尽管总体上来说是有利于生命的，但是，我们只能在一个非常有限的光域内看见宇宙，实际上，几乎完全被限制在可见光和相当有限的无线电波范围之内。我们必须在大气层以外进行观察，来研究不能穿透地球大气层的紫外线、X射线、红外线和无线电波。借助于在离地球200英里高空运行的人造卫星上的仪器，我们就可以进行这种观察。我们不知道这些仪器会记录些什么，如果知道的话，就不值得找这么多麻烦了。但是这个天文学研究领域有待于以后几代人进行研究。

这只是科学研究中诸多新的可能性之一。在研究地球外部大气层、气象学和地球与行星间的空间等过程中，火箭运载工具的发展开辟了这些新的可能性。在被称为空间探索的工作中，有四种主要运载工具。首先是垂直探测火箭。它可以上升到1,000英里或更高的高度，但在200英里以下最有效。这些火箭只是上升到轨道最高处，然后回落到地球。其次，我们拥有绕地球椭圆形轨道运行的人造卫星。卫星至少穿过150英里左右以外的空间。把某物体作为卫星发射必须要有每小时高达18,000英里的速度。如果卫星轨道特别长，超过地球半径（4,000英里）数倍时，我们就采用外层空间探测器。发射速度越快，其返回地球前穿透空间的力量就越大。最后，当速度达到每小时25,000英里时，探测器再也不能返回地球而是成为太阳的卫星，一颗人造卫星。探测器可以经过专门导向从月球附近掠过，也可撞击月球或成为月球的卫星，这就是月球探测器。近来我们已经有了许多这类探测器的实例。

从科学家的观点来看，所有这些运载工具都很有价值。人造卫星的出现并没有使垂直探测火箭过时，外层空间探测器和月球探测器与人造卫星的关系更是如此。在科学家看来，任何一次特定发射的价值都在于试验成功与否，而不只是离开地球的距离。科学家也不关心这些运载工具是否载人，因为仪器可以自动操作并将其记录的数据转换成无线电信号，从数百万英里以外送回到地球。

太空探索简介（Introduction to Space Exploration）

太空探索是指通过太空旅行对地球所处宇宙的性质所进行的探索。自古以来，人们就梦想着离开我们的星球，探索外部世界。20世纪下半叶，梦想成为现实。1957年第一颗人造卫星的发射拉开了太空时代的序幕。1961年人类首次进入太空。从那以后，太空探索持续时间更长，美、俄等国宇航员甚至连续数月生活在环绕空间站飞行的飞船上。共有24人完成绕月飞行或在月球表面行走。同时，机器人探索器到达了人类无法前往的地方，太阳系的主要星球只剩下一处还未留下足迹。无人驾驶宇宙飞船也已经到达过许多小天体，如月球、彗星和小行星。从火箭到通讯设备再到计算机，这些探索活动促进了新技术的发展。宇宙飞船的研究得到了大量关于太阳系、银河系和宇宙的科学发现，并且为人类了解地球及相邻星球提供了新的视野。

太空探索的第一个难题是要研制出足够强大和可靠的火箭，将卫星送进轨道。但这些推进器所需的不仅是蛮力，也需要导航系统使其依照合适的飞行路径，到达预定轨道。太空探索的第二个难题是制造卫星本身。这些卫星的电子元件要求重量很轻，但也要足以承受发射时的推进和振动。要制造这些电子元件，就需要世界范围内的太空工程设施在制造和测试的可靠性上采用新的标准。地球上，工程师还需要建造卫星跟踪站，以便在这些人造"月球"绕地球运转时维持无线电通讯。

20世纪60年代早期，人类开始发射探测器，探索其他星球。这些自动探测器的运行距离使得其运行时间要以数月或数年为单位来衡量。这些太空飞行器要有足够的耐受度，持续运转十年甚至更久。它们还需要应对各种危险情况，如环绕木星的辐射带，环绕土星光环运动的颗粒，同时和地球周边太空飞行器相比，要面对更加极端的温度。尽管这些飞行任务的科学收益很高，它们的成本也很高。如今世界上的航天局，如美国国家航空航天局和欧洲航天局，正在尝试执行更加便宜和高效的机器人探测任务。

在这些无人驾驶飞船进入太空后,人类也会来到太空。载人太空飞行面临着一系列新的挑战,其中很多都和如何在太空的恶劣环境中维持生命有关。由于太空的真空状态,载人宇宙飞船需要自带空气。此外,还有其他致命的危险,如太阳和宇宙的辐射;可能会刺穿宇宙飞船船身或戳破宇航员的压力服的微陨石群(小的岩石和尘埃);从寒冷的黑夜到火辣辣的阳光直射引起的极冷到骤热的温度变化;等等。仅仅使得人类在太空中存活是不够的——宇航员还需要在太空中有效完成工作的方法。也需要开发太空导航和科学观测及实验需要的工具和技术。宇航员离开增压的飞船去真空中作业时也需要受到保护。为了保障从起飞到降落中任何可预见的紧急情况下全体飞船人员的安全,需要事先精心设计任务细节和硬件环境。

绕地球飞行的载人太空飞行任务十分艰巨,将宇航员送往月球的阿波罗计划则更加令人生畏,而将宇航员送到月球表面并成功返回的壮举则标志了人类太空飞行的巅峰。

阿波罗计划之后,载人太空飞行任务的重心转向了更长时间的太空飞行,苏联和美国的空间站率先做出尝试。研制可循环利用的太空飞船是另一个目标,这使美国航空飞船得到飞跃性的发展。目前,科学家们主要致力于两个方面的工作:一是在执行长达一年或更长时间(需要到达附近行星的时间)的太空任务时,保持宇航员的健康状态;二是降低发送卫星到轨道的成本。

Unit Ten   Civil Aviation

Part A   Lecture

Quiz

1.(1) 擅长特技飞行的昆虫

新技术使科学家们能够详细研究昆虫如何飞行以及它们翅翼的构造。这对于开发新的空气动力学技术是有益处的。蜻蜓在飞行中是如何旋转和俯冲的?苍蝇又是如何大回环翻筋斗、仰面朝天地旋转以及笔直地横滚的?昆虫的飞行特技确实令人赞叹不已。可是,这些昆虫在飞行方面是怎样胜过人类的飞行器的呢？

现代技术正在开始解答这些问题。通过高速摄影和电子显微镜扫描研究，发现昆虫擅长飞行的秘密之一在于它们翅翼的设计。

构成昆虫翅翼的膜既薄又坚韧。鸟和蝙蝠是借助于其翅膀里面的关节和肌肉飞行的。然而，昆虫得以在空中飞翔是借助于躯干中的肌肉的运动来带动翅翼进行的。昆虫翅翼的形状可以做细微的变化，这不仅借助于它们肌肉的力量，也借助于它们身上坚硬而柔韧的翅脉的组合，使之得以产生不同程度的弹性。

（2）水中高压脉冲放电的光辐射研究

摘要：（目的）研究水中高压脉冲放电的光辐射性质。（方法）通过高温计示波器系统，测量放电的光谱辐射亮度；采用灰体辐射模型，计算放电通道的温度和发射率；结合放电的电特性，分析这一过程中等离子体的光辐射特性。（结果）在电容器储能为 312.5 J、最大峰值电流为 50 KA. 的条件下，计算得到的最高温度约为 5×10^4 K。光辐射能量的绝大部分分布在远紫外的范围（10 nm—200 nm）。（结论）水中放电的光辐射特性对应于放电的电特性，紫外光辐射是水中高压脉冲放电的主要释能方式之一。

（3）新的半导体

科学家们正在研究用新的方法制造用于计算机和其他现代电子仪器的半导体材料。

目前，半导体是由元素硅制成的。但是科学家们认为，其他材料的效果将会更好。他们说，他们正在研制的材料传输电信号比硅更快。这些材料会发光。这意味着，微小的电系统可以用光来取代电线进行通讯。而且，新的材料不会轻易被辐射、高温或低温损坏，因此可以用于太空卫星和火箭。

在芝加哥附近的西北大学工程师们研制了一种称为磷砷化铟的半导体材料，这是一种铟、磷和砷的混合物。工程师布鲁斯·韦塞尔斯说，使用这种材料的计算机其运算速度将比用硅的计算机快三到六倍。他还说，很容易在实验室或工厂里制造这种材料。

制造磷砷化铟所需的材料成本昂贵，约为硅成本的 100 倍。但是韦塞尔斯教授说，材料费用只占总成本的一小部分。他希望，使用他研制的这种材料的计算机系统将在五年内问世。

另一种有可能替代硅的材料名叫砷化镓，它传送电信号比硅快 5 倍，并且会发光。然而，砷化镓系统价格昂贵，易于损坏，散热也不好。

伊利诺伊大学的科学家们最近说，他们已经研究出一种方法，可以在一个微小的电系统，即芯片上同时使用砷化镓和硅。科学家们说，这样一种结合的系统将具有两种材料的优点。它将像砷化镓一样速度快，能发光，可以抗辐射，又像硅一样，价格也不贵。还可以制成大块。伊利诺伊小组的负责人哈迪斯·莫科克说，他的技术用在一些控制电流的晶体管装置上效果很好。

许多在美国其他大学和公司，以及在日本公司工作的科学家也在研究如何在电系统中使用这两种材料。

Part B Reading

Exercises

1. (1) aviation industry
   (2) fuel efficiency
   (3) carbon-fibre composite
   (4) global warming
   (5) International Air Transport Association
   (6) Paris Air Show
   (7) zero-emission
   (8) greenhouse-gas emission
   (9) Intergovernmental Panel on Climate Change
   (10) fuel bill
   (11) large civilian airliner
   (12) aeroengine
   (13) turbofan
   (14) gas-turbine engine
   (15) budget airline
   (16) open rotor
   (17) geared turbofan
   (18) rotational speed
   (19) Pratt & Whitney
   (20) carbon nanotube
   (21) wind tunnel
   (22) stealth bomber
   (23) blended wing-body design
   (24) net carbon emission
   (25) superconducting generator
   (26) electric motor

2.（1）它的燃油效率提高了20%，并且是第一架主要结构和外壳由轻质碳纤维复合材料而不是金属合金组成的大型客机。

（2）波音公司和空客公司共同制造了世界上大多数大型民用客机，它们都希望在航空发动机方面有所改进。

（3）当今的民用飞机使用涡轮风扇发动机，在其内部，发动机后部的燃气轮机驱动安装在前面的风扇，该风扇可提供大部分推力。

（4）易捷航空建议将涡轮风扇安装在飞机的上方和后方，这样机尾和机身就能阻挡地面发出的噪音，而折断的叶片也不会击中机舱。

（5）预计明年投入商用的波音787在这方面将处于领先地位，其整个机身均采用碳纤维复合材料。

（6）这促使工程师可以在计算机上彻底审查复杂的新设计，然后着手建造成本更高的用于风洞测试的原型。

（7）波音公司和美国宇航局目前正在进行地面测试的混合翼身设计有望减少20%的燃料。

Part C Extended Reading

Exercises

1. (1) air transport　　　　　　　(2) legal issue
　　(3) air route　　　　　　　　　(4) super-sonic aircraft
　　(5) air cargo container　　　　　(6) total loss
　　(7) air traffic control system　　　(8) air carrier
　　(9) aeronautical act　　　　　　(10) international convention
　　(11) international air law　　　　(12) civilian aircraft
　　(13) international agreement　　(14) air service
　　(15) flight schedule　　　　　　(16) airspace

2.（1）运输量的迅速增加致使国内外飞机飞行次数增加。

（2）随着全球航空客货运量的快速增长，以及飞机运行技术的高速发展，世界正逐渐成为"一日生活圈"。

（3）在大陆法系国家，航空法或民商法规定了国内航空运输中旅客与货物托运人的权利和义务以及承运人的责任。

（4）提供服务的航空公司和航空承运人的民用飞机在跨越国际边界进入另一个或多个国家的领空时，或多或少地会受到有关国家政府的管制。

（5）另一方面，有些国家与国家组织一起参与区域管制，尽管参与的大多是不具约束力的共同管制。

参考译文

民航面临环保挑战（Civil Aviation Faces Green Challenge）

作为快速增长的温室气体排放体，航空业承受着提高燃油效率的巨大压力。库尔特·克莱纳就航空业的选择进行了调查。

上周日，波音公司推出了它们的最新客机波音787。它的燃油效率提高了20%，并且是第一架主要结构和外壳由轻质碳纤维复合材料而不是金属合金组成的大型客机。但由于航空业在全球变暖中所扮演的角色，它们面临着压力并正努力地进一步提高效率。

国际航空运输协会发言人安东尼·康西尔说："我们的目标是在环境责任方面将航空业提升到一个新的高度。"

在上个月的巴黎航展上，这一努力成为焦点。空客公司呼吁就未来的技术挑战进行全球合作。国际航空运输协会总干事乔凡尼·比西尼亚尼表示，到2050年应该有"零排放"飞机。

据政府间气候变化专门委员会称，民用航空温室气体排放约占总排放量的3.5%；但随着航空旅行的普及，这一数字可能会上升。鉴于油价高，航空公司的燃油费用开支也推动了对更高的燃油效率的追求。空客公司环境事务总监雷纳·冯·怀德说："我们知道，为了能够将新飞机推向市场，它的油耗必须比将要更换的飞机节省15%至25%。"

波音公司和空客公司共同制造了世界上大多数大型民用客机，它们都希望在航空发动机方面有所改进。当今的民用飞机使用涡轮风扇发动机，在其内部，发动机后部的燃气轮机驱动安装在前面的风扇，该风扇可提供大部分推力。涡轮风扇发动机比纯燃气涡轮发动机更安静、更节能，而纯燃气涡轮发动机的推力完全来自从发动机后部喷出的热空气。如果这些风扇做的大一些，效率就会提高。但随着它们的体积增大，环绕它们的管道体积也随之增大，它们增加的重量和阻力很快就抵消了更大风扇的优势。

在巴黎揭晓的 EcoJet 提案中，英国廉价航空公司易捷航空提议的一种飞机，到 2015 年每客位英里的二氧化碳排放量将减半。该提议采用开式转子，消除风道。位于英国德比的发动机制造商劳斯莱斯的技术开发主管大卫·克拉克表示，更大的风扇可以使产生的发动机效率提高 15%。但是，消除风道会增加噪音，而且在叶片损坏的情况下，对机舱的保护也会减少。易捷航空建议将涡轮风扇安装在飞机的上方和后方，这样机尾和机身就能阻挡地面发出的噪音，而折断的叶片也不会击中机舱。

齿轮传动涡轮风扇发动机也能实现类似的效率提升。在传统的涡轮风扇发动机中，发动机后部的燃气轮机通过轴直接转动风扇。虽然风扇的效率最高，但是转速却比涡轮的最有效转速低得多。传统发动机被迫以折衷的速度运转。

位于康涅狄格州哈特福德的喷气发动机制造商普惠公司已经在测试一种新发动机，该发动机将变速箱置于涡轮机和风扇之间。该公司的下一代产品负责人鲍勃·萨亚表示，它可使风扇以涡轮机转速的三分之一运转，并将效率提高 15%。

除了发动机，飞机制造商还希望改善机身设计以提高燃油效率。最简单的改进是使现有的机身设计更轻。预计明年投入商用的波音 787 将在这方面处于领先地位，其整个机身均采用碳纤维复合材料。人们普遍认为，下一步是采用由碳纳米管增强的更坚固的复合材料。

工程师还利用计算机模拟工具进一步减少飞机的阻力。波音公司在华盛顿州西雅图环境绩效战略总监比尔·格洛弗说，最新版本的波音 737 机翼产生的阻力比它们所替换的机翼少 30%。加利福尼亚州圣地亚哥特可普国际公司的航空分析师汉斯·韦伯补充说，功能越来越强大的仿真工具有望实现更好的设计。这促使工程师可以在计算机上彻底审查复杂的新设计，然后着手建造成本更高的用于风洞测试的原型。

此外，制造商还采用全新的机身概念，例如在美国空军 B2 隐形轰炸机中首创的混合翼身。这些设计消除管状机身和机尾，使整个飞机变成一个升力面。波音公司和美国宇航局目前正在进行地面测试的混合翼身设计有望减少 20%

的燃料。但行业观察家不确定乘客是否会接受这样的设计：他们将不得不坐在一个宽阔的、无窗的隔间里，每当飞机倾斜时，机翼末端附近的乘客就会上下摇晃。机场的登机口也必须完全重新设计以接纳这种飞机。

航空业还在研究新燃料如何帮助其缓解环境问题。生物燃料具有明显的吸引力——只要公众接受，其增长和燃烧不会产生净碳排放。生物乙醇是汽车发动机中最常用的生物燃料，在高海拔地区会结冰，但生物柴油和生物丁醇都可以在传统涡轮风扇发动机中燃烧。总部位于伦敦的维珍航空公司董事长理查德·布兰森承诺，将在2008年使用生物燃料商用飞机，并与波音公司和通用电气的航空发动机部门合作，测试候选燃料。

最终的零排放燃料可能被证明是氢，它可以很容易地被传统的喷气式发动机燃烧而不产生碳排放。首先，我们必须弄清楚如何经济而安全地制造和分离氢。 即便如此，飞机也需要重新设计。尽管液态氢提供的能量是同等重量煤油的三倍，但它占的空间却是煤油的七倍。这需要飞机配备更大的油箱以及保温材料以保持液体冷却。

美国宇航局格伦研究中心(位于俄亥俄州克利夫兰)的物理学家杰拉尔德·布朗说，一旦飞机具备了可以使用液态氢作为燃料的冷却设备，就有可能制造出更高效的全电动飞机。传统的发电机太重，无法在飞机上高效使用。但他认为，有了冷氢源，超导发电机和电动机就开始有意义了，并且它们可能比现在的涡轮风扇发动机更有效。

最后，航空业将协商一致，寻求将所有这些方法结合起来的办法，以避免人们指责其快速增长正在刺激全球变暖。波音公司的格洛弗说："我们有很多有创意的人，有很多很棒的想法。我想说，没有什么想法是不可以谈的。"

国际公共航空法简介
(A Brief Introduction to the International Public Air Law)

航空运输在当今世界的重要性，特别是各国之间航空运输量的日益增加，以及经济和高科技的发展，使我们律师对各自国家的相关法律问题自然而然地产生了好奇心。运输量的迅速增加致使国内外飞机飞行次数增加。因此，在某些航线上，飞机飞行的频率已经过高。

随着全球航空客货运量的快速增长，以及飞机运行技术的高速发展，世界正逐渐成为"一日生活圈"。于是，许多国家的航空公司纷纷拓展海外航线，增加飞机航班，以吸引更多的乘客和货物，因此各国之间的竞争更加激烈。

特别是自从超音速飞机和航空货运集装箱出现以来，世界各地的航空客运和货运量逐年增加。同时，随着飞机客货运业务的增加，飞机事故发生的频率也越来越高。

亚洲、欧洲和美洲发达国家以及发展中国家都有很多航空案例。飞机事故造成损害的特征有：（1）重大损害；（2）全损（全有或全无）；（3）即时损害（瞬间）；（4）与地面有从属关系损害（空中交通管制系统）；（5）国际损害。这些飞机事故在确定赔偿责任限额或不限额方面，以及进行损害鉴定时，都会引起受害人与航空承运人之间的诸多争议。

也许更重要的是，为保证乘客安全、飞机有序快速飞行以及解决受害人与航空承运人之间的争端，他们必须尽职地遵守与航空运输有关的法规、行政程序、航空条例和国际公约。

有关客运和货运的国际航空法是以国际条约、议定书和公约为基础的。

在国际航空运输早期，世界上没有统一的法律法规来管理旅客或货物的运输。

在大陆法系国家，航空法或民商法规定了国内航空运输中旅客与货物托运人的权利和义务以及承运人的责任。提供服务的航空公司和航空承运人的民用飞机在跨越国际边界进入另一个或多个国家的领空时，或多或少地会受到有关国家政府的管制。

在大多数情况下，由两个或两个以上主权国家管制的每次飞行，都需要在有关国家之间就如何开展其商业航空服务达成正式的国际协议。此外，各国分别对航空服务的各个方面进行监管，例如向本国航空承运人发执照、要求其主管部门填写关税和航班时刻表、收集统计数据、保护消费者等。另一方面，有些国家与国家组织一起参与区域管制，尽管参与的大多是不具约束力的共同管制。国际民用航空在法律上受国家对其领空主权原则的管辖。

Unit Eleven　Computer Science

Part A　Lecture

Quiz

1.（1）药品营养信息

复合维生素 B 是几种维生素 B 的化合物，它能促进食物转化为能量，并保持神经系统的正常工作。本产品能 100% 满足美国日摄入量参考标准（RDI）中规定的叶酸日摄入量。研究表明，叶酸有益于保持成年男女的心血管系统正常运行。妇女在健康饮食中摄入足量的叶酸能降低有先天性大脑或脊髓缺陷的婴儿的出生率。

本药品药片整体包有薄膜衣，易于吞咽。

（2）磁带使用说明

　　松下 VHS 盒式录像带
　　只适用于 VHS 录像机

如何使用及保存

使用

- 为防止意外抹去录下的资料，VHS 盒式录像带在背面设有一个可拆除的小护片。
- 小护片可用螺丝刀扳掉。一旦这个护片被扳掉，录像带只能用于回放。
- 如果想用没有护片的录像带进行录像，只要用胶带将护片处的洞盖住即可。

保存

- 存放之前，记住要将录像带放回盒中，并竖直放置。
- 存放于阴凉干燥处，避免阳光直射或靠近热源。
- 避免录像带跌落，也不要使其受到冲击。
- 最好将录像带全部倒好后再存放。
- 如果录像带倒得不平整，重新倒一遍，使其绕整齐。

- VHS 盒式录像带不得粘接，所以不要试图修理或拆卸录像带。
- 不要触摸带子的表面。您皮肤上的污垢或油脂会损坏带子表面的磁膜。

（3）电视机保养须知

- 本机灵敏度极高，务必小心轻放。
- 本机不得放在暖气管、炉灶等热源旁边和满布灰尘的地方，也不得置于机器震动、冲撞之处。
- 本机不得放在阳光直射之处或潮湿的地方。
- 注意本机四周应保持通风，以防机内积热，无处散发。
- 勿把本机放在地毯、毛毯等柔软物体上，或靠近窗帘、帷幕处，以免空气流动受到阻塞。
- 本机内装有危险的高电压，使用时切勿打开机箱，以防触电。
- 保存装放本机的厚纸箱和捆包填塞物，便于日后需要重新包装运送时使用。捆包机器时，最好按照出厂时的原捆包形式。
- 为了保持机器外表的崭新，应时常用柔软的抹布擦拭。
- 如遇不易脱落的污秽，可用抹布沾上一点中性洗涤剂轻轻擦拭。切勿使用稀释剂或苯等烈性溶剂或摩擦粉，以免机体外部受损。
- 若长期不使用，应拔下墙上插座的插头。拔电线时，应拉住插头拔出，切勿抽拉电线，以免折断。

Part B Reading

Exercises

1. (1) motor control
   (2) driverless car
   (3) Internet of Thing
   (4) sensor
   (5) sat-nav system
   (6) search engine
   (7) movement on the stock market
   (8) virtual reality
   (9) military drone
   (10) robot minesweeper
   (11) life science
   (12) biological evolution
   (13) human intelligence

2.（1）但所有这些都涉及心理技能，如感知、联想、预测、规划和运动控制等；这些技能使人类和动物能够实现其目标。
（2）金融家用来预测股市走势的系统以及国家政府用来指导卫生和交通政策决策的系统也同样如此。
（3）不那么令人高兴的是，军用无人机在当今的战场上穿梭；但值得庆幸的是，机器人扫雷机也这样做了。
（4）仍然有许多悬而未决的问题，因为人工智能本身告诉我们，我们的大脑比心理学家以前想象的要丰富得多。
（5）他们用这些来解决，如众人皆知的身心问题、自由意志的难题以及许多关于意识的困惑。
（6）最后，同样重要的是，人工智能已经挑战了我们思考人类及其未来的方式。的确，有些人担心我们是否真的有未来，因为他们预见到人工智能将全面超越人类智慧。

Part C Extended Reading

Exercises

1. (1) cloud technology　　(2) nervous system
   (3) word-processing　　(4) payroll
   (5) inventory　　(6) paperless office
   (7) customer interaction　　(8) product delivery
   (9) inflow of data　　(10) outflow of data
   (11) data flow　　(12) financial transaction
   (13) data volume　　(14) social media
   (15) data ecosystem　　(16) feedback loop
   (17) instant notification　　(18) data silo
   (19) cloud computing　　(20) smart agent
   (21) systems thinking　　(22) machine learning

2.（1）在社交、移动设备和云技术的推动下，一个重要的转变正在发生，这将引领我们进入当今这些公司所处的数据驱动世界。

（2）具有数字神经系统的组织特征是大量数据的流入和流出、内部和外部的高度网络化、增加的数据流以及随之而来的复杂性。

（3）当其他人过渡到数字本地操作时，为处理大型网络公司获取的互连异构数据而开发的技术将成为我们的主要工具。

（4）我们发现了这样的早期实例，从捕获金融交易中的欺诈行为到人力资源部门的调试和改进招聘流程：几乎每个人都已经注意到与他们相关的大量社交网络信息的流动。

（5）传感器和机器人技术为工业、运输和军事等领域带来当今网络公司所拥有的优势，也许世界上最大的转变就要来了。

（6）海量数据流的挑战以及层次结构和边界的侵蚀，将引导我们使用统计方法、系统思维和机器学习，以应对我们正在创造的未来。

参考译文

人工智能（Artificial Intelligence）

人工智能试图让计算机做大脑能做的事情。

其中一些能力（如推理）通常被描述为"智能"，而其他的能力（如视觉）则不是。但所有这些都涉及心理技能，如感知、联想、预测、计划和运动控制等；这些技能使人类和动物能够实现其目标。

智能不是一个单一的维度，而是一个具有多种信息处理能力的、结构丰富的空间。因此，人工智能使用许多不同的技术，解决许多不同的任务。

而且它无处不在。

人工智能可实际应用于家庭、汽车（和无人驾驶汽车）、办公室、银行、医院、天空以及互联网，包括物联网（物联网将越来越多的物理传感器连接到我们的小装置、衣服和环境中）。还有一些人工智能在我们的星球之外：被送到月球和火星的机器人，或在太空轨道上运行的卫星。好莱坞动画、视频和电脑游戏，卫星导航系统以及谷歌的搜索引擎都是基于人工智能技术的。金融家用来预测股市走势的系统以及国家政府用来指导卫生和交通政策决策的系统也同样如此。手机上的应用程序也是如此。在虚拟现实中添加化身，以及为"伴侣"机器人开发的"试水阶段"情感模型。甚至连美术馆都在自

Key and Reference for Translation

己的网站上使用人工智能，也在电脑艺术展览中使用人工智能。不那么令人高兴的是，军用无人机在当今的战场上穿梭；但值得庆幸的是，扫雷机器人也这样做了。

人工智能有两个主要目标。一个是技术性的：使用计算机完成有用的事情（有时采用的方法与头脑中使用的方法非常不同）。另一个是科学的：使用人工智能的概念和模型来帮助回答有关人类和其他生物的问题。大多数人工智能工作者只关注其中之一，但有些人会同时考虑两者。

除了提供无数的技术小发明，人工智能还对生命科学产生了深远的影响。特别是，人工智能使心理学家和神经科学家能够开发出强大的思维—大脑理论。这些模型包括物理大脑如何工作的模型，以及解决一个不同但同样重要的问题，即大脑在做什么：它在回答什么计算（心理）问题，以及什么样的信息处理使它能够这样做。仍然有许多悬而未决的问题，因为人工智能本身告诉我们，我们的大脑比心理学家以前想象的要丰富得多。

生物学家也以"人工生命"的形式使用人工智能，开发出生物体不同方面的计算机模型。这有助于他们解释各种类型的动物行为、身体形态的发展、生物进化以及生命的本质。

除了影响生命科学，人工智能还影响了哲学。如今，许多哲学家把他们的思想描述建立在人工智能概念上。他们用这些来解决，如众人皆知的身心问题、自由意志的难题以及许多关于意识的困惑。但是，这些哲学思想存在着极大的争议。对于是否有人工智能系统能拥有真正的智能、创造力或生命，人们的意见存在着极大的分歧。

最后，同样重要的是，人工智能已经挑战了我们思考人类及其未来的方式。的确，有些人担心我们是否真的有未来，因为他们预见到人工智能将全面超越人类智慧。虽然有一些思想家对这种前景表示欢迎，但大多数人对此感到恐惧：他们问，人类的尊严和责任是否还有一席之地？

大数据（Big Data）

"大数据"里所有的数据从何而来？为什么大数据不只是脸书和谷歌等公司关心的问题？答案是，网络公司是先行者。在社交、移动设备和云技术的推动下，一个重要的转变正在发生，这将引领我们进入当今这些公司所处的数据驱动世界。

从外骨骼到神经系统

直到几年前，计算机系统在社会，特别是商业中的主要功能还是作为一个数字支持系统。应用程序对现有的现实世界流程进行了数字化处理，如文字处理、工资单和库存。这些系统通过商店、人员、电话、船运等方式将界面返回到现实世界。"无纸化办公"这个现在很奇怪的短语暗指将已有的纸质流程转移到计算机中。这些计算机系统形成了一个数字外骨骼，为现实世界中的企业提供支持。

互联网和万维网的出现增加了一个新的维度，带来了一个完全数字化的商业时代。客户交互、付款和通常的产品交付完全可以在计算机系统中发生。数据不再仅仅停留在外骨骼中，而成为操作中的关键要素。我们正处在一个商业和社会都在走进数字神经系统的时代。

具有数字神经系统的组织特征是大量数据的流入和流出、内部和外部的高度网络化、增加的数据流以及随之而来的复杂性。

这种转变正是大数据重要的原因。当其他人过渡到数字本地操作时，为处理大型网络公司获取的互连异构数据而开发的技术将成为我们的主要工具。我们发现了这样的早期实例，从捕获金融交易中的欺诈行为到人力资源部门的调试和改进招聘流程：几乎每个人都已经注意到与他们相关的大量社交网络信息的流动。

跟踪转变过程

随着技术在商务领域中的进展，每一步都会带来数据量的飞跃。对于现在看大数据的人来说，一个合乎情理的问题是，他们的业务不是谷歌或脸书，为什么大数据适用于他们？

答案在于网络企业能够100％进行在线活动。从开始一直延伸到结束，他们的数字神经系统很容易操作。如果你有工厂、商店和现实世界的其他企业，则需要进一步将它们融入数字神经系统。

但"进一步计划"并不意味着它不会发生。网络、社交媒体、移动设备和云端的驱动力正将更多的业务带入数据驱动的世界。在英国，政府数字服务部门正在统一向公民提供服务。结果是公民体验的彻底改善，且许多部门第一次能够真实了解自己的工作情况。对于任何零售商来说，像Square、American Express和Foursquare等公司，都将付款方式带入一个社会化、反应灵敏的数据生态系统中，从而将信息从企业会计的筒仓中解放出来。

拥有数字神经系统意味着什么？关键特征是使组织的反馈回路完全数字化。也就是说，从传感和监控输入直接连接到产品输出。这在网上很简单。零售业变得越来越容易。传感器和机器人技术为工业、运输和军事等领域带来当今网络公司所拥有的优势，也许世界上最大的转变就要来了。

过去的 30 年里，数字神经系统的覆盖范围稳步增长，每一步都带来了敏捷性和灵活性的提高，同时，重要的是，也带来了更多的数据。首先，从特定的应用程序到使用个人电脑的一般业务应用程序。然后，通过网络直接交互。移动设备增加了对时间和地点的感知以及即时通知。下一步，进入云端，打破数据孤岛，通过云计算增加存储和计算弹性。现在，我们正在整合智能代理，让它们代表我们行事，并通过传感器和自动化连接到现实世界。

来了，准备好了没有

如果你没有考虑更多的数字化运营优势，那么，准没错，你的竞争对手正在考虑。正如马克·安德森去年所写的："软件正在吞噬整个世界。"一切都变得可编程化。

正是这种数字神经系统的发展，使得大数据的技术和工具与今天的我们息息相关。海量数据流的挑战以及层次结构和边界的侵蚀，将引导我们使用统计方法、系统思维和机器学习，以应对我们正在创造的未来。

Unit Twelve　Mechanical Engineering

Part B　Reading

Exercises

1. (1) mechanical engineering　　(2) civil engineering
   (3) mining engineering　　(4) metallurgical engineering
   (5) chemical engineering　　(6) electrical engineering
   (7) structural engineering　　(8) hydraulic engineering
   (9) sanitary engineering　　(10) environmental engineering

(11) industrial engineering　　(12) nuclear engineering
(13) petroleum engineering　　(14) aerospace engineering
(15) electronic engineering

2. (1) theoretical knowledge　　(2) quantification
   (3) explosion of scientific knowledge　(4) experimental method
   (5) water supply　　(6) purification
   (7) sewer system　　(8) project organizer
   (9) specialty　　(10) on-the-job training

3. （1）工程业是历史上最古老的行业之一。要是没有广阔的工程学领域里所拥有的种种技术的话，那么，绝不可能有今天的文明。

   （2）人们常常把工程学定义为实际应用理论科学（如物理学和力学）的学科。工程学初期的许多分科都不是以科学为基础，而是以经验资料为基础的；而后者有赖于观察结果与经验，而非理论知识。

   （3）因此，数学是现代工程学的语言，这样说怎么都不算太过分。

   （4）现在机械工程师有能力用数学方法来计算许多种不同机械之间复杂的相互作用所产生的机械效益。

   （5）专门职业是指像法律、医学或工程学那样的要求受到专业高等教育的行业；事实上，它们常被称为"有学问的职业/行业"。

Part C　Extended Reading

Exercises

1. (1) dynamics　　(2) automatic control
   (3) thermodynamics　　(4) fluid flow
   (5) heat transfer　　(6) lubrication
   (7) properties of material　　(8) development
   (9) operation　　(10) maintenance
   (11) maximum value　　(12) minimum investment
   (13) management　　(14) consulting

(15) marketing
(16) operations research
(17) value engineering
(18) PABLA (problem analysis by logical approach)

2. （1）第一个作用是理解和研究机械学科的基础。

（2）这一工作不仅要求对机械学科有一个清楚的了解，并具有把复杂系统分解成基本因素的能力，而且还需要有综合和发明的创新性。

（3）其目的在于维护或提高企业或机构的长期生存能力和声誉的同时，以最少的投资和成本产生最大的价值。

（4）在所有这些作用中，体现出一种长期不断地使用科学的方法，而不是传统的或直觉的方法的倾向，这是机械工程专门化水平不断提高的一个方面。

（5）正如在其他领域一样，在机械工程中，能够迈出创新的重要一步去开创新方法，在很大程度上仍然是个人的、即兴的能力。

参考译文

工程行业（The Engineering Profession）

工程业是历史上最古老的行业之一。要是没有广阔的工程学领域里所拥有的种种技术的话，那么，绝不可能有今天的文明。那些用石块削制箭头和枪矛的初期工具制作者，是现代机械工程师的先驱。那些在地球上发现了金属并找到了精炼和使用金属的办法的工匠，是采矿和冶金工程师的祖宗。而那些设计灌溉系统并建造古代神奇建筑物的技术娴熟的技工，则是当时的土木工程师。从历史上流传下来的最古老的名字之一，是伊霍德普，他是大约公元前三千年古埃及撒喀拉阶梯形金字塔的设计者。

人们常常把工程学定义为实际应用理论科学（如物理学和力学）的学科。工程学初期的许多分科都不是以科学为基础，而是以经验资料为基础的；而后者有赖于观察结果与经验，而非理论知识。那些当时想出办法把建造英国巨石阵和世界独有的埃及金字塔所需要的巨石劈开的人，在试验中发现了楔的原理，因为金字塔很可能是在建筑的过程中利用环绕在四周的斜土坡逐步堆砌上去的；这是斜面原理的一个实际应用，尽管人们在当时对斜面原理的概念还不能从量上去了解，也不能用数学公式去表示。

　　量化是自 16 世纪和 17 世纪现代化时代开始以来造成科学知识爆炸的主要原因之一。另一个重要因素则是用来检验理论的实验方法获得了发展。量化涉及把根据实验获得的数据或各种零星资料归纳成精确的数学表达式。因此，数学是现代工程学的语言，这样说怎么都不算太过分。

　　自 19 世纪以来，数学成果在科学上和实际问题上的应用都有了逐步的发展。现在机械工程师有能力用数学方法来计算许多种不同机械之间复杂的相互作用所产生的机械效益。人们可以使用新型的、强度更高的材料，并且有了庞大的新能源。工业革命从驱使水和蒸汽工作开始。从此之后，用电、用汽油和用其他能源的机器已非常普及，它们现在干着世界上绝大部分的工作。

　　科学知识迅速扩展的一个结果，是科学专业与工程专业两者在数量上的增长。到 19 世纪末，不但产生了机械工程、土木工程以及采矿和冶金工程，而且还出现了化学工程与电气工程这些新专业。这种扩展过程一直延续至今。例如，我们现在有核工程、石油工程、航空航天工程以及电子工程等。在每一工程领域内还有细分的专业。例如，在土木工程领域里，就有下列这些分科：结构工程——研究永久性结构；水力工程——涉及水或其他流体的流动与控制的系统；以及环境卫生或环保工程——研究供水系统、水净化系统与排污系统等。机械工程的主要分支是工业工程，它涉及工业上的全套机械系统，而不是研究单台机器。

　　科学知识增长的另一个结果是，工程学已发展成为一种专门职业 / 行业。专门职业是指像法律、医学或工程学那样的要求受到专业高等教育的行业；事实上，它们常被称为"有学问的职业 / 行业"。19 世纪以前，工程师通常还只是些技工或工程项目的组织者，他们通过学徒、岗位培训或反复试验等方法学会技艺。而现在，许多工程师要花费多年时间在大学里为高等学位而学习。然而，即使那些不是为了高等学位才学习的工程师们，也必须了解自己领域里的发展变化及与自己领域有关的发展变化。

　　这样，工程师这个词在英语里就有两种用法。其一是指专业工程师，他们有大学学位，学习过数学、科学和某一门工程专业。但工程师还用来指操作或维护发动机或机器的人。一个很贴切的例子便是开火车的铁路机车工程师。这个意义上的工程师实质上是技术员，而不是专业工程师。

Key and Reference for Translation

机械工程的作用（Functions of Mechanical Engineering）

机械工程有四个作用。第一个作用是理解和研究机械学科的基础。它包括涉及力和运动之间关系的动力学，如在振动中的力和运动的关系；自动控制；研究各种形式的热、能量、动力之间关系的热力学；流体流动；热传递；润滑以及材料特性。

第二个作用是依次地进行研究、设计和开发。这个作用试图带来必要的变化以满足当前和将来的需要。这一工作不仅要求对机械学科有一个清楚的了解，并具有把复杂系统分解成基本因素的能力，而且还需要有综合和发明的创新性。

第三个作用是生产产品和动力，包括计划、运作和维护。其目的在于维护或提高企业或机构的长期生存能力和声誉的同时，以最少的投资和成本产生最大的价值。

第四个作用是机械工程的协调作用，包括管理、咨询、以及在某些情况下进行市场营销。

在所有这些作用中，体现出一种长期不断地使用科学的方法，而不是传统的或直觉的方法的倾向，这是机械工程专门化水平不断提高的一个方面。这些新的合理化方法的典型名称有：运筹学、工程经济学、逻辑法问题分析（简称 PABLA）。然而，创造性是无法合理化的。正如在其他领域一样，在机械工程中，能够迈出创新的重要一步去开创新方法，在很大程度上仍然是个人的、即兴的能力。